OH NO, NOT POWERPOINT! RELAX, HELP IS ON THE WAY.

A POWERPOINT REPAIR MANUAL

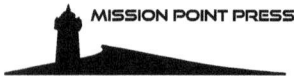

MISSION POINT PRESS

Published by Mission Point Press
2554 Chandler Rd.
Traverse City, MI 49696
(231) 421-9513
www.MissionPointPress.com

Design by Sarah Meiers
*Cover art compiled with imagery from Adobe Stock
by WDnet Studio and ROMAN RYBALKO*

ISBN: 978-1-961302-64-8
Library of Congress Control Number: 2024908903

Printed in the United States of America

OH NO,
NOT POWERPOINT!
RELAX, HELP IS ON THE WAY.

A POWERPOINT REPAIR MANUAL

KEN HASELEY

Author of
Change the Way You Communicate:
Why You Should. How You Can.

MISSION POINT PRESS

Contents

Most PowerPoint presentations miss the mark. This amazing tool is simply being misused. Let's change that!

Introduction

Shortly after taking the reins as CEO of IBM in 1993, Louis Gerstner, Jr. asked for a briefing on the state of the company's mainframe business. According to Gerstner, "At that time, the standard format of any important IBM meeting was a presentation using overhead projectors and graphics on transparencies that IBMers called . . . 'foils.' Nick [Donofrio, who was then running the System/390 business] was on his second foil when I stepped to the table and as politely as I could in front of his team, switched off the projector. After a long moment of awkward silence, I simply said, 'Let's just talk about your business.' By that afternoon an email about my hitting the Off button on the overhead projector was crisscrossing the world."

Today, standard format for presentations at most companies involves PowerPoint. Virtually no important corporate decision takes place without it. Guess what? Turning off the projector usually still makes sense! You probably know why:

- Most presentations contain a seemingly endless number of slides, causing the eyes of audience members to glaze over.
- Those slides are usually crammed full of too much information – lots of bullet points (and sub-bullets), unnecessary words, complete sentences, multiple charts loaded onto a single slide, and the company logo on every slide.
- Slides are poorly designed – with cliché photos, words or images so small they're hard or

impossible to make out, unusual type or color choice, jarring white space, etc.
- Presenters turn their backs to the audience and read what's on the screen.

I could go on . . . and on.

Visual aids can be a powerful tool to help convey information in a way that's clear, creative and compelling. Today, the most popular presentation-graphics tool is Microsoft's PowerPoint. Nearly everyone uses it. The *Wall Street Journal* reports that even second graders are drawn to it. Talk about "show and tell."

Despite PowerPoint's widespread use, perhaps no other communications tool in business gets as much ridicule — from the very people who use it or sit through it.

But despite its widespread use, perhaps no other communications tool in business gets as much ridicule — from the very people who use it or sit through it. "Death by PowerPoint" has become a familiar expression in corporate America.

The fact is most PowerPoint presentations miss the mark. This amazing tool is simply being misused. And it's understandable: few, if any, of us have been shown how to use it effectively. Chances are you never got a PowerPoint tutorial in high school, college, or from your employer. We watch how others in our organization use it and we do likewise. The blind are leading the blind. And yes, some companies provide guidelines for PowerPoint use, but those guidelines are rarely in sync with principles of effective communication.

There are thousands of workshops, books, articles and online tutorials about PowerPoint. Maybe you've attended, read or watched one of them. But those resources focus almost exclusively on what I call a "production objective." They show you how to produce visuals using the amazing features of this innovative software.

But what about the more important "communications objective" – getting the audience to listen to, hear, understand and act on what you say? Most likely you've never been shown how PowerPoint can either help or hinder in that objective.

Which brings us to this book. It examines a variety of ideas – some of them unconventional, many of them new and research-based – about what constitutes effective use of PowerPoint. Much of the information you're about to encounter runs counter to what we regularly see during PowerPoint presentations. So, you're likely to raise some eyebrows or get some pushback when you begin to make changes in how you use PowerPoint in your organization. But don't worry; I'll show you how to stay in your company's good graces while inching forward with incremental improvements that will help you achieve that all-important communications objective.

Consider this book a PowerPoint repair manual. Reading it will change the way you think about and use PowerPoint. Now, let's start making some repairs.

Compared with taste, touch, smell and hearing, vision is the most powerful sense humans have for receiving information.

Chapter 1

The Power of Visual Communication

Visual Aids: What the Research Says

If the presentations you deliver typically include a visual component, such as PowerPoint, a whiteboard or flip chart, maybe a prop – congratulations are in order. Why? Because those visuals could be giving you and your presentations an edge. Let me explain:

Back in 1986, researchers from the University of Minnesota and from 3M Corporation (inventor of Post-It Notes) wanted to know whether visual aids improve presentations. So they conducted a study to test the impact of visual support (e.g., 35mm slides, overhead transparencies, etc.) on presentations designed to persuade. (Persuasion involves influencing attention, comprehension, degree of agreement with the presenter's position, and retention.) That research, along with other research conducted at Harvard, University of Pennsylvania, UCLA and University of Wisconsin, provides some eye-opening insight into visual communications. Consider the following:

- Presentations using visual aids are 43% more persuasive than unaided presentations.
- Audience retention of information can increase 50% when people see as well as hear.

- Visuals improve learning by 200%.
- A speaker's goals are met 30% more often when visuals are incorporated into a presentation.
- Group consensus in meetings occurs 20% more often when visuals are involved.
- Using visual aids can reduce by 40% the time required to present complex ideas.
- Compared with taste, touch, smell and hearing, vision is the most powerful sense humans have for receiving information. Some 70% of the sense receptors in our bodies are dedicated to vision.
- Humans process visuals 60,000 times faster than they process text.
- People who use high-quality color (vs. black and white) visuals are perceived to be more professional presenters.
- Products or services presented with a multi-media component (e.g., slides, sound, video, props, etc.) are more likely to be perceived as credible, professional or reliable — even if they are not.
- Visuals are a way to present complex technical information, emphasize important information, vary the manner of conveying information, and appeal to different learning styles.

One interesting take on visual communication comes from two customer-research specialists. In a *Harvard Business Review* article titled, "Change the Way You Persuade," Gary A. Williams and Robert B. Miller point out that, "People have a natural tendency toward a certain style of decision making that gets reinforced through success — or that changes after repeated failures." That default style places them in one of five categories — one of which is Charismatics.

As the name implies, Charismatics are enthusiastic, high-energy individuals. They love new ideas. According to the authors, think: Richard Branson and Oprah Winfrey, to name a few examples. (Steve Jobs and Elon Musk would be on my list.) Charismatics tend to process the world visually, so it's best to use visual aids when trying to persuade them about the value of an idea or proposal. Is your boss or prospective customer a Charismatic? If so, don't forget PowerPoint!

Powerful stuff – the above. But to take advantage of all this firepower, make sure your presentation visuals are well crafted and well delivered. Read on.

A Brief History of PowerPoint

The idea for PowerPoint first surfaced in 1984. Robert Gaskins, a computer scientist at a software company called Forethought, was looking for an alternative to overhead transparencies and chalkboard presentations. He and two developers, Thomas Rudkin and Dennis Austin, spent several years creating a presentation-graphics program – initially for the Apple Macintosh computer. Apple was so impressed with what was being developed, it selected Forethought for its first-ever venture capital investment – $432,000. The program being developed was initially called Presenter, but because that name was already trademarked, Presenter was renamed PowerPoint. (Apple specified that "PowerPoint" be one word with the second "P" in caps so the name would be consistent with the names of all Macintosh software applications.)

The first version of PowerPoint for Macintosh was designed so black and white pages could be printed, then photocopied onto overhead transparencies. Printing speaker notes and audience handouts was the other purpose. PowerPoint 1.0 was an immediate success; the first production run of 10,000 units sold out.

Microsoft took note and scrapped plans to develop its own presentation-graphics software. Instead, it purchased

The Night Before the Big Meeting Frank Receives a Visit from the PowerPoint Fairy.

Forethought for $14 million in 1987. (Initially, Microsoft CEO Bill Gates was skeptical about acquiring PowerPoint, but eventually signed off on the acquisition.)

In 1990, PowerPoint was released for Windows PC. By that time, it included color and a wide variety of other features, including spell check, find and replace, and shading. PowerPoint, which started out as an alternative to overhead transparencies and 35mm slides, was fast becoming their replacement.

OH NO, NOT POWERPOINT! RELAX, HELP IS ON THE WAY.

In succeeding years, PowerPoint continued to evolve, offering users an impressive array of features, including: presentation templates, sound and video, rehearsal mode, hidden slides and presentation view (tools such as notes, thumbnails, time clock, etc. – visible only to the presenter during the slideshow).

Estimates are that PowerPoint holds approximately a 95% share of the presentation software market.

Today, PowerPoint is available in more than 100 languages. Estimates are that it holds approximately a 95 percent share of the presentation software market – far surpassing competitors such as Keynote, Prezi and Google Slides.

Although it's the 800-pound gorilla of presentation software apps, PowerPoint is not the only game in town.

Chapter 2

PowerPoint Alternatives (Past and Present)

PowerPoint Competitors

Although it's the 800-pound gorilla of presentation software apps, PowerPoint is not the only game in town. Here are four others:

Prezi (www.prezi.com)

Prezi is a Hungarian visual and video communication software company founded in 2009. (The name Prezi is a shortened version of the Hungarian word for "presentation.") The company is best known for its unique, storytelling software alternative to traditional slide-based presentation formats such as PowerPoint.

Traditional presentations have slides organized in a linear sequence – slide after slide, usually with abrupt transitions between them. Prezi's approach provides users with a canvas or map – an overview that identifies the topic and its components or sub-topics that can be accessed in any order – sequentially or randomly. Click on a sub-topic to zoom in on or out of its details.

For example, with Prezi, the topic, *Creating a Top-flight Presentation* might have five components: Preparation, Visuals, Practice, Delivery and Q&A. These terms would appear together on screen, and the presenter could select each one in that order or in any order, and then drill down for details.

Prezi presentations look different. They stand out, incorporate movement and have some sizzle. (Some might say they are too theatrical or gimmicky.) Also, with Prezi, presenters can incorporate videos of themselves narrating some or all of the presentation – which can be archived and viewed by others at another time.

Prezi offers a free, basic-level (but limited) product allowing you to create and share up to five visual projects. Other, more comprehensive products involve a cost per person per year.

Keynote (www.apple.com/keynote)

Keynote is Apple's presentation software application. It began as a computer program for Apple CEO Steve Jobs who used it to create presentations for various conferences and events. It became available to the public in 2003 and is currently available in 33 languages.

Anyone familiar with Apple products knows that they are well designed, free of extraneous features and easy to use. Keynote is no different. Users can create first-rate visual presentations in a few steps. The software takes a minimalist design approach; visuals are simple, clear, devoid of "gingerbread." There are fewer tools available on Keynote than on PowerPoint, but there are some impressive features. For example, Keynote allows you do such things as create presentations with multiple speakers in different locations and add live video feeds to any slide.

Keynote is free to Apple users and is available on Mac, iPad, iPhone and through iCloud on the web. It is not available on Windows and Android devices, so quite a few people will not be able to use it. (It is possible to convert Keynote files into PowerPoint, but some animations and graphics may not work properly.)

Google Slides (www.google.com/slides)
Google Slides is a web-based presentations tool that can be accessed anytime, anywhere – if you have internet access. There's no app to download. The software is free and available to anyone with a Google account.

PowerPoint users will find this slideshow tool easy to use. It offers a wide range of templates and themes, along with a number of free add-ons. There's even a separate tutorial for users who are weaning themselves off PowerPoint. However, most people will be able to master Google Slides in about a half hour.

Where Slides stands out from most of its competitors is in its sharing and collaboration capabilities. You're able to share your slide deck with others, and you control the permission level – letting them view only, comment only or edit the presentation. Up to 100 users with editing permission can make changes simultaneously, and these changes will be instantly visible on every device that's connected to the file.

Canva (www.canva.com)
The "new kid" on the online graphic design block, Canva is an Australian company founded in 2013. Its stated mission is to help people create professional-quality designs for presentations, social media, websites, logos, video and more.

For business and other presentations, Canva's software

app offers thousands of free templates – significantly more than are offered by its competitors. The company owns Pixabay and Pexel, two free stock photo sites, so users have access to millions of photos, icons, graphics, audio clips and sound effects.

Canva's presentation software allows users to collaborate, edit and present from anywhere on any browser or mobile device. Several noteworthy features:

- Magic Design: Uses AI to generate a first draft (outline, slides, content) of your presentation.
- Talking Presentation: Allows you to practice, record and share a video for audiences to watch from anywhere at any time.
- Magic Shortcuts: Lets you surprise your audience with a drumroll, confetti or a mic drop. (Caution: clever, but a potential gimmick.)

Cost: Canva offers three product packages: Canva Free, Canva Pro (access to premium content) and Canva for Teams (designed to allow teams to collaborate).

Overhead Projectors

If you've never seen or used an overhead projector, it's a piece of equipment that projects images from letter-size plastic film. A bright light shines through the film onto a mirror, which reflects the image through a lens and then onto a screen.

Overhead projectors were used in U.S. military training as early as 1940 and by the late 1950s and early 1960s they were widely used in schools and business. Educators liked how the device promoted interactive learning. Teaching materials could be pre-printed on plastic sheets, and the instructor could write directly on those sheets with a

35mm Slides (R.I.P.)

Remember the 1980s movie *Gung Ho* starring Michael Keaton? It's about a shuttered auto plant in Pennsylvania in search of a new owner.

In the movie, Keaton's character travels to Tokyo to try to persuade a Japanese auto firm to reopen the factory. In one scene, Keaton arrives at the firm's headquarters carrying a slide projector, carousel tray and portable screen. As he enters the boardroom, he asks where he should set up the screen. Then he sees that the room already has a first-rate A-V setup, prompting him to tell the dour-faced executives in the room (sarcastically), "Let me just set this down. Don't really need it . . . after I lugged it about 14,000 miles! Alright, don't worry about it."

It wasn't too long ago that slide presentations were delivered using a projector, carousel tray and film mounted in 35mm slides. In some ways the slides could be considered works of art. They were created by graphic artists – by hand. Trained artists painstakingly cut and pasted various elements such as pictures, drawings and type, creating a single image that was photographed onto 35mm film. These works of art took time and skill to create and were expensive – usually hundreds of dollars per slide.

Slides were loaded into narrow slots on a projector's carousel tray, and the presenter would forward the tray (usually with a wired remote control), one position at a time – causing the slide to drop into a gate in front of the light source which projected the image onto a large screen in the room. Depending on the projector, moving from slide to slide (forward or reverse) usually took between 0.9 to 1.5 seconds. Forwarding or reversing the tray created a distinct, noticeable clicking sound.

As primitive as this technology may now seem, it had some advantages: Perhaps most importantly, slides were created by people who understood design – color, typeface, proportion, etc. It was their full-time job. Contrast that to PowerPoint, which puts design decisions in the hands of non-experts – and the results are often disastrous. Also, because of how they were made, slides tended to have much less information, especially significantly less text.

On the flip side, slides took time (and money) to produce – typically several days. And last-minute revisions (a given with presentations by executives) usually incurred 100- to 150-percent rush charges. Other problems included slides being loaded backwards or upside down into the carousel tray, slides getting jammed in the projector, and notoriously short-lived projector bulbs (not easy to change during a presentation).

35mm slide projectors came into widespread use during the 1950s. Their production ended around 2004. Kodachrome film, which was used to make the slides (and according to a song by American singer-songwriter Paul Simon, gave us "those nice, bright colors,") was discontinued in 2009.

washable color marking pen. The projector could be placed at a comfortable writing height, and the teacher could face the students and interact more effectively with them. No need to turn your back to the class and write awkwardly in large script on a chalkboard.

These projectors use high-power halogen lamps that generate a lot of heat which needs to be cooled – usually by a noisy fan. This intense heat often causes bulbs to burn out in less than one hundred hours (compared to modern LCD or DLP projector lamps which last for thousands of hours).

Moving an overhead projector requires some muscle. Units tend to be heavy and bulky and are usually moved around and used on wheeled carts. Compared to ceiling- or

In the movie *Apollo 13*, flight director Gene Kranz (played by Ed Harris) tries to sketch something on an overhead transparency in a critical ground crew meeting. He wants his team to find a way to get the astronauts stranded 205,000 miles from Earth in a crippled spacecraft back home. Just as he turns on the projector, the bulb blows, so he goes low-tech and uses the room's chalkboard instead.

OH NO, NOT POWERPOINT! RELAX, HELP IS ON THE WAY.

table-mounted LCD projectors, they take up a lot of room space.

This technology is still around, but in the 2000s it began to be replaced by computer projection systems which allow users to move more easily from image to image and show animations and video. Also, today's presenters see no reason to tolerate the low-quality images of these projectors (e.g., too bright in the center, too dim around the edges).

Blackboards

The blackboard probably got its start as an individual writing slate for students in 11th-century India. By the 16th century, larger blackboards were being used in Europe – primarily in music education. In the United States, it wasn't until about 1800 that a math teacher at West Point began to use connected slates to share complex formulas to large classes.

Today, although black slate has given way to green, porcelain-enameled steel, and chalk has been replaced by gypsum, blackboards remain a prominent feature in some classrooms, especially in developing

countries. Blackboards are inexpensive, durable, and help teachers slow down their instruction so students can more easily digest the information.

Compared with their extensive use in education, blackboards were not widely used in business – except perhaps in retail. Today, they're making a comeback. Bars, restaurants, coffee shops, grocery stores like the "old school" feel and personal touch they convey for menus and other messages. Blackboards and colored chalk are giving some of those cold, high-tech monitors a run for their money.

Whiteboards

True story: Back in the 1980s I was working for an energy and chemical company that was growing through acquisitions. The CEO had set his sights on acquiring a major railroad. Yes, a railroad! It turns out a lot of railroad companies have access to vast tracts (no pun intended) of land. And the land this particular railroad controlled contained extensive coal and other energy reserves. The acquisition made sense.

So the CEO assembled his board of directors and top management in the boardroom to explain the rationale behind the proposed acquisition. During the meeting, the room's whiteboard filled up with a treasure-trove of sensitive information – name of the target railroad, financial projections, acquisition timetable and code name, etc.

When the meeting ended, the person responsible for erasing all that confidential information realized that it had been written with a permanent marker. The room was promptly locked and a security guard stood watch until someone could find some isopropyl alcohol and eliminate every mark on the board, including the readable remnants of what was wiped away.

Permanent marker mistakes aside, whiteboards have

become a popular communications tool in schools and business. These blackboard offspring are found in virtually every conference room and in many classrooms and individual offices.

Some presentations lend themselves to real-time collaboration between presenter and audience. For example, if a company is considering a new product launch, a brainstorming session among the right people can help in the decision-making process. Or, after a crisis, a crisis management team can benefit from a post-crisis briefing, where correct actions, missteps and needed improvements can be identified. Two-way communication promotes creativity and learning. And whiteboards promote two-way communication.

In situations where the content of a presentation is flexible or even open-ended – where multiple perspectives are important – whiteboards work well. They also make sense for informal presentations, when the need for a presentation arises unexpectedly or when audience make-up might change at the last minute.

Bonus: Whiteboards can provide leaders or aspiring leaders who know how to use them with instant credibility and can convey executive presence. How so? Thinking (and writing) quickly on your feet is an impressive skill.

Tips:
- Conference room whiteboards can be quite expansive – filling an entire wall or two. Use that real estate wisely. Instead of writing content on just any open space, organize it – perhaps sequentially or in categories, columns or sections as appropriate. Colored, dry-erase markers can help you avoid disorganization. Also, as the session progresses, erase any content that's no longer needed.
- If your session involves co-presenters, share the writing duties. If you're the sole presenter, there might even be a situation where it makes sense to call on an audience member (alerted in advance, of course) to help you out at the board. The audience will welcome this interactive change of pace.
- Write clearly (with non-permanent markers). Printing is better than cursive writing.
- When using a whiteboard, don't "wing it" – putting information on it willy-nilly. Remember this line from a Rod Stewart song? "Her ad-libbed lines were well rehearsed." Have a plan for using the whiteboard but make how you use it seem spontaneous.

Smart Whiteboards

If you've ever watched TV news coverage of local or national election results, you probably saw someone who covers politics standing in front of a large screen, tapping portions of it to call up various kinds of data. That screen was a smart whiteboard (also known as a digital, electronic or interactive whiteboard).

Introduced around 1990, smart whiteboards are large, computer-integrated or computer-connected, high-resolution display screens that project images, video and other multi-media content. Presenters control what's displayed by using their finger on a touch-sensitive screen, similar to touch screens on tablets or smart phones. These whiteboards, some as large as 86 inches, can be wall-mounted or used on moveable stands. They're perfect for presentations delivered in boardrooms or conference rooms. Remote audience members can connect and interact through their smart phones or tablets.

The technology is impressive. Among other things, it allows your audience to collaborate in real time. Features such as freehand drawing tend to boost creativity. Plus, you'll never run out of space to share or archive your content; smart whiteboards are cloud-based — providing an infinite canvas with unlimited storage. And perhaps most importantly, presentations delivered via smart whiteboards are different and interesting; they fascinate and engage audiences.

Drawbacks? Smart whiteboards can be expensive; fifty-five-inch boards can easily cost $5,000 or more. There's also the potential for audience distraction, where the technology (rather than your content) takes center stage. And here's a big one: Because the screen is the control panel, presenters tend to stay close to it to activate the next image. In other words — too much facing the screen; not enough eye contact with the audience. (We'll discuss the "85/15" rule a bit later.)

Flip Charts

They may be leaning against a back wall in some conference room or stashed in a storage closet, but flip charts are still around . . . . and are useful. They can add visual and vocal energy to a presentation. Writing or drawing on a flip chart tends to energize a speaker through greater movement and more voice volume and inflection. That, in turn, can energize an audience. What's more, using a flip chart gives you control over the flow of information; writing as you present keeps the audience from reading and thinking ahead.

Flip charts are well suited for use with smaller audiences. For example, in a sales call on one or two people, using PowerPoint — whether projected on a large screen or displayed on a laptop — might seem a bit awkward. Flip charts can even be used with larger audiences if you have the video technology to project your information onto a large screen.

If one goal of your presentation is to generate greater audience involvement, flip charts can help. There's something about them that prompts more questions and discussion. That's why brainstorming sessions usually involve a flip chart. And if you're conducting a public meeting — especially one where the audience is skeptical, angry or even hostile, flip charts can prevent or minimize outbursts. How? Use the chart to record audience concerns or comments. When an audience sees you writing those comments down, they know you are listening, and they are less likely to act out in socially unacceptable ways.

When presenting, don't feel you need to choose between a flip chart and PowerPoint. Use both. It adds variety. Let's say your company decided to change its name and logo. At a company meeting where the new name is being revealed to employees, have the speaker briefly transition from using PowerPoint (or perhaps just talking about the reason for the change) to using a flip chart. Have him or her write some of the names that were considered, but rejected, and comment on each. Next, write the name that was selected, and elaborate on it. Then, return to PowerPoint — showing the new company logo, and continue with the presentation. Sure, it's a performance, but one most executives can pull off. . . with a bit of practice.

Some do's and don'ts:
- Check the pad's paper supply.
- Be sure everyone in the audience can see the chart.
- Use dark-color markers (e.g., black, blue, etc.); avoid pastels.
- Print (large) rather than use cursive writing.
- Use single words or short phrases; avoid complete sentences.
- Don't feel compelled to write and talk at the same time. Audiences welcome brief periods of silence to process (or enjoy) what you've said; also, the silence provides the speaker with some additional think time.

Props

Short for "theatrical properties," props are movable objects used by stage or screen actors during a performance. Props add a touch of reality to a story. In the film *Field of Dreams*, characters played by Kevin Costner and James Earl Jones go to Fenway Park to catch a ballgame. At the concession stand, they each order a "dog and a beer." The cups placed on the counter prominently display the Miller Lite logo. Companies often pay big bucks to get their products used as props.

In a presentation, props can grab audience attention and turn an abstraction into something concrete. They're memorable. But they're not widely or frequently used anymore, which is a shame because they can have real impact. (The fact that few business presenters are using them makes them even more valuable; they've not lost their power through overuse.)

If you decide to use a prop, keep the following in mind:

- Make sure it relates to one of your key points. As shown in a nearby sidebar, squeezing lime juice into a glass during a presentation helped drive home the point that technology exists that can extract crude oil from a "depleted" oilfield. Props are shortcuts that grab audience attention and facilitate understanding and retention.
- Know how to use the prop. Typically, that includes usually concealing the object until the right moment. (In the TED Talk described in an adjacent sidebar, Bill Gates had a jar of mosquitos on stage in full view of the audience before he used it. But it wasn't a distraction and may have even intrigued the audience.) Also, be sure to include the prop in your practice. Using a prop is a performance, and most of us are not born performers.
- Be sure your audience can see the prop. This can

At the 2008 Macworld Conference & Expo, Steve Jobs introduced the MacBook Air, which at the time was the thinnest notebook computer. After describing some of its features, Jobs said the computer was "so thin it even fits inside one of those envelopes you see floating around the office." Then he walked to the side of the stage, picked up an inter-office envelope, untied its closure, opened the envelope and pulled out the computer. The audience roared its approval!

OH NO, NOT POWERPOINT! RELAX, HELP IS ON THE WAY.

be a problem with larger venues and audiences – unless you have access to large-screen video projection of the speaker.

Finally, some props make great leave-behinds. One of my clients is an oilfield services company that held one of its analysts' meetings at its research center. The event included displays showing some of the new technology being developed to find and produce oil and gas. In his presentation, the research VP referenced hockey great Wayne Gretzky and reminded the audience of Gretzky's famous quote, "I skate to where the puck is going to be, not to where it has been." The exec then went on to say that the developing technology on display would pay dividends down the road by addressing future challenges faced by oil and gas companies. Using that powerful analogy led the company to give a hockey puck with the company's name and logo on it to each analyst. Those props probably ended up as paperweights – a constant reminder of the company's forward-thinking approach to research. (Who says communication has to be boring?)

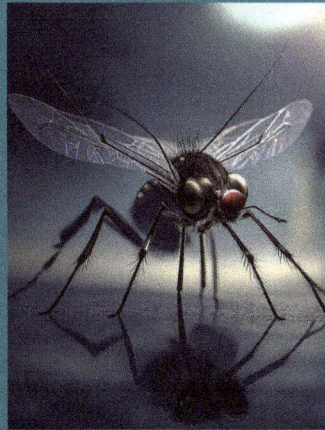

In a TED Talk on malaria, Bill Gates pointed out that the million deaths a year caused by malaria greatly underestimate its impact. More than 200 million people at a time suffer from it, which means that you can't get the economies in these afflicted areas going.

At one point, as he said, "Now, malaria is of course transmitted by mosquitos," he moved toward a small container of mosquitos sitting on a table on stage, opened it and continued, "I brought some here so you could experience this. We'll let those roam around the auditorium a little bit . . . There's no reason only poor people should have the experience . . . Those mosquitos are not infected." The audience laughter was infectious (pardon the pun). Gates' prop and the message connected to it were memorable.

I teach a communications course in the University of Houston's Executive MBA program. To provide students with practice in crafting powerful presentations, I have them form teams and I give each team a scenario. The team's mission is to develop one part of a presentation based on that scenario — for instance: the opening, the closing, a compelling argument or example, etc. One member of the team then delivers that part of the presentation to the class.

One team's presentation was a fictional company's "road show" seeking funding from angel investors. The company specialized in recovering oil and gas from fields that were thought to be depleted.

OH NO, NOT POWERPOINT! RELAX, HELP IS ON THE WAY.

The presenter began by talking about the standard process of exploring for and producing oil and gas – namely, that the reserves that are easiest and cheapest to tap are targeted first. As he spoke, he took a lime, cut it in half and began squeezing out the juice by hand into a glass. Naturally, he could extract only so much liquid.

Then he continued his discussion of oil and gas reserves. As attractive reserves are depleted, and as demand for energy (and its price) rises, there's an incentive to revisit those "depleted" fields or target other less attractive fields. His fictional company had the technology and the expertise to get at those reserves.

He picked up that squeezed lime half, put it into a small, hand-held press, held it over the glass, and squeezed the handles – producing even more juice.

Give that team of students an "A." They understood how props can be used to engage an audience and effectively convey an idea – in this case, the connection between squeezing a lime and extracting oil. In fact, the next time the students in that class see a lime or a juice press, they'll probably think of oil recovery.

What we sometimes forget is how powerful and effective communicating through the spoken word – without any visuals – can be.

Chapter 3

Hold On, Let's Take a Second Look at PowerPoint

Shocking News: Not Every Presentation Needs PowerPoint

I'm a University of Houston adjunct faculty member who developed and teaches Communications for Leaders, the first course students take in their Executive MBA program. Some years back, the dean told me that a team of students complained to him about an unsatisfactory grade they got on a presentation they delivered in a marketing course. The professor penalized them for not using PowerPoint. The students explained to the professor they were taught that using PowerPoint was not always necessary.

Many people in business, government and the professions believe that PowerPoint must be a component of every presentation. They are wrong.

Because that bit of advice came from me, the dean asked if I would provide the professor with my perspective on PowerPoint. I jumped at the opportunity to do some evangelizing and provided him with some thoughts about the use (or overuse) of PowerPoint. Now, not having seen the students' presentation, I have no idea whether the team should have incorporated PowerPoint. Perhaps so. What I do know is this: many people in business, government and the professions (including that colleague of mine in academia) believe that PowerPoint must be a component of every presentation. They are wrong.

Without a doubt, PowerPoint can play an important role in presenting. For example, if you want to identify what share of a particular market your company and five of your competitors have, using a bar graph or multi-colored pie chart showing percentages is probably the best way to communicate that information. In this book, we'll talk at length about effective use of PowerPoint. (I promise.) But first, let me provide a shorter version of the perspective I shared with that marketing professor.

PowerPoint use in presentations is ubiquitous. One reason is that many people feel that credible, professional presentations cannot be created without it. And besides, doesn't everybody use it? Another reason has to do with delivery: those PowerPoint slides projected on a large screen serve as a convenient prompt for the presenter. (PowerPoint can be a comforting security blanket.) There may even be some people who are so uncomfortable speaking in public they would rather have the audience focus its attention on the visuals (in a darkened room) rather than on the presenter.

What we sometimes forget is how powerful and effective communicating through the spoken word – without any visuals – can be. Think about some of the sermons, eulogies or commencement or inaugural addresses you've heard. Or maybe you've had what National Public Radio calls a "driveway moment." While driving, you're listening to a particular story on NPR; it's so compelling that even after pulling into your driveway or garage, you continue listening in the car until the story concludes. Don't underestimate the power of PowerPoint-free communication.

PowerPoint excels at conveying information through charts, photos, tables and to a lesser extent, words. But remember, no visual can compete with the power of a human being to connect with an audience. Before you can reach someone on an intellectual level, you must first connect with them on an emotional level. Powerful presenters rely on themselves rather than on projected images to reach an audience. Develop the confidence and competence to rely on yourself when your messages do not lend themselves to visual representation.

One more point: It comes from Roy Underhill, author of *Khrushchev's Shoe and Other Ways to Captivate an Audience of 1 to 1,000*.

According to Underhill, audiences have certain expectations for production values in different kinds of communications. In film, television, print media and stage productions, the audience expects high production values. For example, in the documentary, *An Inconvenient Truth*, watch Al Gore climb onto a hydraulic crane with a railed platform. It raises him up to the top of the room's projection screen filled with data about how high carbon dioxide levels on Earth have risen. Entertaining . . . and effective.

In one-on-one or in small- or medium-sized group communication, audiences have lower production value expectations.

In short, develop a well crafted presentation, but when incorporating visuals, factor in audience expectations. Satisfy but don't overreach.

A PowerPoint Parody

Abraham Lincoln's Gettysburg Address is a mere ten sentences – 271 words, and Lincoln delivered it in just over two minutes. You'll recall that the purpose of the speech was to dedicate the Gettysburg cemetery and eulogize the 50,000 fallen soldiers. Eulogists at that time traditionally spoke for hours, but Lincoln was so concise that photographers were still setting up their equipment as he finished. That's why there are no photos of him delivering the speech. In spite (or because) of its brevity, the address is a giant among historical American speeches.

Peter Norvig, a computer scientist and sometime critic of PowerPoint, wanted to join the ranks of those who believe presentations are often damaged by unnecessary visuals. So he created a hilarious parody that imagined Lincoln using PowerPoint at that cemetery in Pennsylvania. Here's Norvig's graphic and written versions of the Gettysburg Address:

Good morning. Just a second while I get this connection to work. Do I press this button here? Function F7? No, that's not right. Hmmm. Maybe I'll have to reboot. Hold on a minute. Um, my name is Abe Lincoln and I'm your president. While we're waiting, I want to thank Judge David Wills, chairman of the committee supervising the dedication of the Gettysburg Cemetery. It's great to be here, Dave, and you and your committee are doing a great job. Gee, sometimes this new technology does have glitches, but we couldn't live without it, could we? Oh – is it ready? OK, here we go:

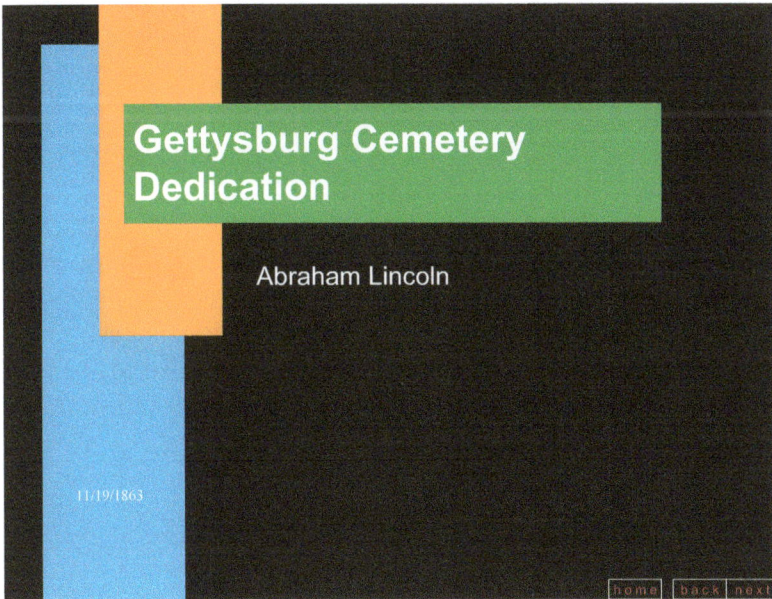

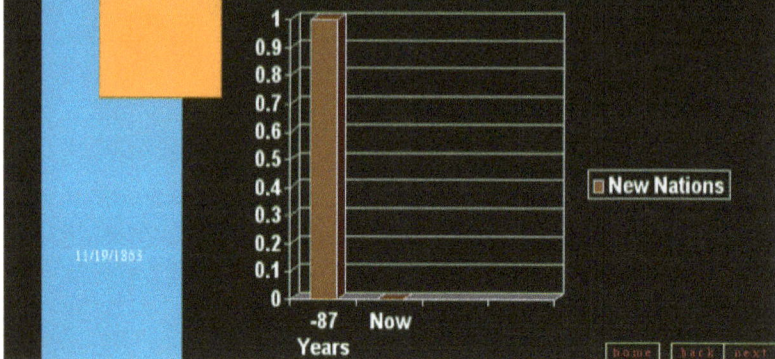

As we see in the Organizational Overview slide, four score and seven years ago our fathers brought forth on this continent a new nation, conceived in liberty and dedicated to the proposition that all men are created equal. Now we are engaged in a great civil war, testing whether that nation or any nation so conceived and so dedicated can long endure. Next slide please.

We are met on a great battlefield of that war. We have come to dedicate a portion of that field as a final resting place for those who here gave their lives that that nation might live. It is altogether fitting and proper that we should do this.

But in a larger sense, we cannot dedicate, we cannot consecrate, we cannot hallow this ground. The brave men, living and dead who struggled here have consecrated it far above our poor power to add or detract. The world will little note nor long remember what we say here, but it can never forget what they did here.

Review of Key Objectives & Critical Success Factors

- What makes nation unique
  - Conceived in Liberty
  - Men are equal
- Shared vision
  - New birth of freedom
  - Gov't of/for/by the people

11/19/1863

home   back   next

It is for us the living rather to be dedicated here to the unfinished work which they who fought here have thus far so nobly advanced. It is rather for us to be here dedicated to the great task remaining before us – that from these honored dead we take increased devotion to that cause for which they gave the last full measure of devotion . . . that we here highly resolve that these dead shall not have died in vain, that this nation under God shall have a new birth of freedom, and that government of the people, by the people, for the people shall not perish from the earth.

Of course, Lincoln couldn't have used PowerPoint at Gettysburg. He predated the technology by 124 years. And I think most of us would agree: that was a good thing. Before using PowerPoint, ask yourself whether it will enhance your presentation or detract from it.

PowerPoint's Role in the Space Shuttle Disaster

In 2003, the Space Shuttle Columbia disintegrated as it re-entered the Earth's atmosphere, killing all seven crew members. Shortly after the disaster, the Columbia Accident Investigation Board studied the accident in detail and concluded that it was caused when the shuttle's left wing was hit by insulating foam that had come loose during launch.

But the board also pointed a finger at another culprit: PowerPoint. Investigators argued that in its briefings, NASA relied too heavily on PowerPoint rather than on traditional written reports. For example, when discussing possible wing damage during a mission, engineers used an incredibly complex PowerPoint slide – loaded with bullet points, sub-points and jargon. "It is easy to understand how a senior manager might read this PowerPoint slide and not realize that it addresses a life-threatening situation," the report stated.

Here's that slide (actually, one of many offending slides) – delivered in a presentation prior to the Columbia disaster:

Review of Test Data Indicates Conservatism for Tile Penetration

---

- **The existing SOFI on tile test data used to create Crater was reviewed along with STS-87 Southwest Research data**

  -- **Crater overpredicted penetration of tile coating significantly**

    - **Initial penetration to described by normal velocity**

      - Varies with volume/mass of projectile (e.g. 200ft/sec for 3cu. In)

    - **Significant energy is required for the softer SOFI particle to penetrate the relatively hard tile coating**

      - Test results do show that it is possible at sufficient mass and velocity

    - **Conversely, once tile is penetrated SOFI can cause significant damage**

      - Minor variations in total energy (above penetration level) can cause significant tile damage

      -- **Flight condition is significantly outside of test database**

      - **Volume of ramp is 1920cu in vs. 3 cu in for test**

BOEING   2/21/03   6

The first thing you notice are words – lots of them – 127 to be specific. The audience is overwhelmed (and turned off) by this information overload and will begin either reading or scanning the content – perhaps while the presenter is saying something critical.

Nancy Duarte, who wrote a thoughtful book called *slide:ology: The Art and Science of Creating Great Presentations*, would say that this slide is not a slide at all. "If a slide contains more than 75 words, it has become a document . . . True presentations focus on the presenter and the visionary ideas and concepts they want to communicate. The slides reinforce the content visually rather than create distraction, allowing the audience to comfortably focus on both. It takes an investment of time on the part of the presenter to develop and rehearse this type of content, but the results are worth it."

The slide is bullet- and sub-bullet heavy.

The final bullet, buried at the bottom of the slide, in small type and overshadowed by other data, indicates that the foam insulation tested prior to the mission was more than 600 times smaller than the size of a chunk of insulation that could cause damage in actual flight conditions. It would be easy for the audience to miss or downplay the significance of what might have been the most important piece of information on the slide.

Note that the slide contains jargon: SOFI, Crater, ramp. In this case, the specialized language was probably not a problem for the likely technically trained audience. But it's worth remembering that in every presentation, all jargon must be explained at the first reference – even when the audience may be familiar with it. No exceptions.

Finally, it appears no proofreading took place. One bullet reads, "Initial penetration to described by normal velocity" ???

And Then There's Edward Tufte

No discussion of PowerPoint would be complete without the thoughts of Edward Tufte. A statistician and professor emeritus at Yale University, Tufte is recognized as perhaps the world's leading expert on information design – how information is communicated visually.

In 2003, Tufte published a booklet denouncing PowerPoint titled, "The Cognitive Style of PowerPoint." As he sees it, "Rather than providing information, PowerPoint allows speakers to pretend that they are giving a real talk, and audiences to pretend that they are listening." In the booklet, Tufte lists what he considers to be PowerPoint's many flaws. Among them:

- PowerPoint is "presenter-oriented" not "audience-oriented." Its focus is on making presenting easier, not on ensuring content quality.
- It has a commercial feel (e.g., slides with company logos) that "turns information into a sales pitch and presenters into marketers."
- The software is preoccupied with format over content – linear, slide-by-slide progression; bullet points and sub-bullets; and an artificial, hierarchical structure (like those puzzling rules for outlines: if you have an "A," you must have a "B." If there's a "1," there must be a "2").
- PowerPoint encourages simplistic thinking by asking presenters to summarize sometimes complex ideas in just a few words.
- It breaks information into fragments, so the audience is unable to "see the forest for the trees."
- It lends itself to "chartjunk" – useless information, decoration, gratuitous graphics, etc.
- PowerPoint creates an authoritarian presenter/subordinate audience relationship that discourages or prevents a give-and-take exchange of ideas.

Maybe you think Tufte is onto something. Amazon founder Jeff Bezos certainly does. He sent his senior team an email with this subject line: "No powerpoint [sic] presentations from now on." When his colleagues want to share an idea, they must put it in a four-to-six-page memo. Bezos explained, "The reason writing a 4-page memo is harder than 'writing' a 20-page powerpoint [sic] is because the narrative structure of a good memo forces better thought and better understanding of what's more important than what, and how things are related. Powerpoint-style presentations somehow give permission to gloss over ideas, flatten out any sense of relative importance, and ignore the innerconnectedness of ideas."

If he heard about Amazon's PowerPoint ban, Edward Tufte probably smiled. He advocated something similar when he argued that the most effective way to present information is through a brief, written report read by all participants in the first five to ten minutes of a meeting, followed by discussion and debate.

It's unlikely that the world's most popular tool for presenting information will be going away anytime soon. And though some of Tufte's PowerPoint criticism is compelling, much of it seems extreme – especially given today's desire for speed and brevity. Perhaps Tufte's real contribution lies in getting those of us who use PowerPoint to find ways to overcome its deficiencies – but not abandoning it.

Need Proof of the Power of PowerPoint-free Communications?

Watch some of these videos on YouTube.

Commencement Addresses:

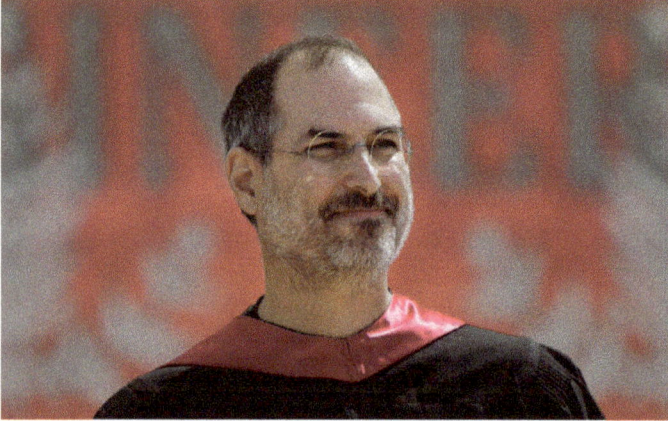

Steve Jobs, Apple CEO
Stanford University, 2005
This speech is believed to be the most-watched commencement address of all time. Jobs delivered the address while battling pancreatic cancer. (*YouTube: Steve Jobs' 2005 Stanford Commencement Address*)

Former Admiral William H. McRaven
University of Texas at Austin, 2014

What's so important about making your bed? Admiral McRaven answers that question in his inspiring twenty-minute speech to the 2014 graduating class of the University of Texas. The speech, titled, "Make Your Bed," went viral on the internet and led McRaven to write a *New York Times* best-selling book with the same title. (*YouTube: Admiral McRaven Addresses the University of Texas at Austin Class of 2014*)

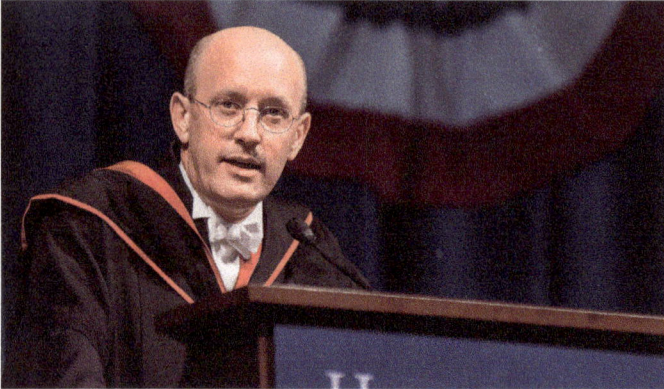

Michael Ward, Senior Research Fellow, University of Oxford
2015 Commencement Address, Hillsdale College

Most commencement addresses disappoint. They are tedious and usually contain a litany of predictable platitudes. Graduating students deserve better. And Ward gave them better. This is one of the most thoughtful, well crafted and entertaining addresses I've come across. You'll especially enjoy the playful opening. (*YouTube: Michael Ward Delivers Hillsdale College's 2015 Commencement Address*)

Eulogy:

Father Paul D. Scalia, Vicar for Clergy, Diocese of Arlington Homily of Christian burial for U.S. Supreme Court Justice Antonin Scalia, February 20, 2016
Father Scalia, a Roman Catholic priest and son of the late Supreme Court Justice Antonin Scalia, delivered a poignant yet humorous eulogy (with an unexpected opening) at his father's funeral. Among the stories he shared was one when his father was in line to make his confession, but left the line when he realized the priest hearing confessions was his son. (*YouTube: Homily of Christian Burial for US Supreme Court Justice Antonin Scalia*)

Inaugural Address:

John F. Kennedy
January 20, 1961
Kennedy's 1,366-word inaugural address is considered one of the best presidential inaugural speeches in American history. It was written by Kennedy and his speechwriter Ted Sorensen. You probably already know the most memorable line in the speech. (*YouTube: President John F. Kennedy's Inaugural Address*)

Special Addresses:

Franklin D. Roosevelt
Pearl Harbor speech to Congress, December 8, 1941

FDR's 6½-minute "Day of Infamy" speech was delivered to a joint session of Congress the day after the American naval base at Pearl Harbor in Hawaii was attacked by Japan. About a half-hour after Roosevelt finished speaking, Congress declared war on Japan. Also, after the speech, military recruiting stations were so inundated with crowds, they had to stay open 24 hours a day to sign up volunteers.

Paul Sparrow, former director at the Franklin D. Roosevelt Presidential Library, says, "This address was not written by a committee of speechwriters and consultants. ... It was dictated by FDR without notes to his assistant Grace Tully just three hours after he learned of the attack. Tully later recalled that he took a long drag on his cigarette, and then 'he began in the same calm tone in which he dictated his mail. Only his diction was a little different and he spoke each word incisively and slowly, carefully specifying each punctuation mark and paragraph.' He dictated the speech ' ... without hesitation, interruption or second thoughts'." (*You Tube: President Franklin D. Roosevelt Declares War on Japan – Full Speech*)

Martin Luther King, Jr.

Address at the March on Washington, August 28, 1963

Who's not familiar with Dr. King's historic "I Have a Dream" speech? But did you know that much of the speech was improvised on the spot? Clarence B. Jones, a former advisor and speechwriter to Dr. King, recalls that at one point when King was reading from the text, his favorite gospel singer Mahalia Jackson, who was on the podium, shouted, "Tell 'em about the dream, Martin, tell 'em about the dream."

According to Jones, King looked at her, then pushed the script to the side of the lectern and looked out at the crowd of more than 250,000 people. Said Jones, "I turned to the person standing next to me . . . and I said, 'These people out there, they don't know it, but they're about ready to go to church'."

The rest of the speech was improvised! Do yourself a favor: watch and listen to this amazing orator give the speech that was not entirely the one he planned to give. (*You Tube: Martin Luther King | I Have a Dream Speech | August 28, 1963, Full Speech*)

Ronald Reagan
Space Shuttle Challenger explosion address to the nation, January 28, 1986

President Reagan postponed the State of the Union speech scheduled for January 28, 1986, so he could address the nation about the space shuttle disaster that occurred that day. Seventy-three seconds into the flight, the shuttle burst into flames, killing all seven crew members on board – among them, teacher Christa McAuliffe, the first civilian to travel into space. Reagan's comforting words, some of them aimed at America's schoolchildren, were delivered in four short minutes. (*YouTube: Challenger: President Reagan's Challenger Disaster Speech – 1/28/86*)

PowerPoint was designed for the audience – to help it get and retain information. But it has turned into a prompt for the presenter.

Chapter 4

Create a Top-flight PowerPoint Presentation

First Things First

For most people, the process of creating a presentation starts with PowerPoint. They begin by thinking about and developing the visuals they plan to use – photos, illustrations, charts, even bullet points. This is the wrong approach. If you're guilty of it, you are putting the cart before the horse. Here's what needs to take place before you consider PowerPoint:

1. Begin with a thorough audience analysis

The more you know about the audience, the more likely you are to make correct inferences about how to reach them.

- Who are they?
- Why are they there? (By choice?)
- What is their knowledge level of your topic?
- What is their interest level in your topic?
- What information needs and wants do they have?
- What size is the audience?
- Are there any audience sensitivities you should be aware of?

Speaking of audience sensitivities, in 1990, First Lady Barbara Bush was asked to deliver the commencement address at Wellesley College, a private, women's liberal arts college in Massachusetts. When students learned who'd be speaking, many of them objected, and voiced their objection. (If you're wondering why – no, it wasn't because Mrs. Bush was

Students at Wellesley College objected to the choice of First Lady Barbara Bush as commencement speaker. Mrs. Bush won them over with a single remark.

a conservative or Republican – although that reason seems reasonable. Students objected because Mrs. Bush wasn't a college graduate, and they saw her as someone whose fame came from the achievements of her husband: ambassador to China, head of the CIA, president of the United States.)

The news media picked up on the brewing controversy: Would the school dis-invite Mrs. Bush? Would the students be rude as she spoke? Or maybe even boycott the commencement? None of that happened. On June 1, Mrs. Bush showed up (and brought along Mrs. Gorbachev, who happened to be in the United States at the time).

At one point in her remarks, Mrs. Bush said, "And who knows? Somewhere out in this audience may even be someone who will one day follow in my footsteps and preside over the White House as the president's spouse. . . . I wish him well!" Thunderous applause followed.

Mrs. Bush knew there were sensitivities in the audience, and she won the students over with that line. It was her way of assuring them that America would one day have a female president.

One of the EMBA courses I teach at the University of Houston meets on Monday and Thursday evenings from 5:30 to 9:30. Nearly all the students in that course are working professionals who have put in a full day of work before class. And for some students, English is their second language. Those are audience sensitivities that impact how I conduct those four-hour sessions.

2. Know the following
- What's the venue? (Formal or informal presentation? Panel discussion?)
- Are there other speakers? (If so, how many and what are their topics?)
- What's your allotted time?
- Will there be a Q&A segment?
- What's the room setup? (If possible, visit the room or get a photo.)

3. Determine the purpose of your presentation

Are you there to inform or educate, demonstrate, inspire, motivate, persuade, entertain? Most presentations combine several of these objectives. Be clear about your purpose; it impacts content.

4. Decide on a central message

When it comes to conveying information, two types of objectives are in play: a production objective (i.e., how you'll deliver the information — for example, through an advertisement, a memo, a speech, a news release, etc.) and a communications objective — the message you want your audience to hear, understand and act on. Focus on the latter. Write your message down in one or two sentences. Anything longer than that is clutter. For example, if someone asks why you shop at Home Depot, you tell him in a single, declarative sentence. Achieve similar brevity when developing your key message. Too many people define success in presenting as simply getting through their presentation (that's a production objective) rather than having their message received and understood (that's a communications objective).

5. Gather information

This is the research phase of crafting. Think about how we buy gifts for people. If a holiday, birthday or some other event is approaching, a little voice inside our head reminds us that dad is an outdoorsman, so be on the lookout for something cool he could use while hunting or fishing. Then, when we're walking through the mall, thumbing through those Sunday paper ad inserts or surfing the web and see something cool, we immediately know it's the perfect gift for dad. Putting on our "dad glasses" enables us to look at merchandise from his perspective. When crafting a presentation, put on your "audience glasses." Doing so will help you determine what information you need to assemble and share.

6. Organize that information into an outline

Remember back in high school or college how important an outline was when you had to write a term paper? It helped you create order out of chaos. Develop the body of your presentation (the longest, most content-rich part) from an outline. Major points or sections of the body become the Roman numerals. Supporting points and detail are the letters and numbers.

> I. Key point
>   A. Subpoint
>   B. Subpoint
>   1. Supporting detail
>   2. Supporting detail
> II. Key point
>   A. Subpoint
>   B. Subpoint

Using this format helps you avoid including information that doesn't fit your purpose or central message and makes it easier to eliminate information if length or time considerations arise.

7. Develop a powerful opening and a strong closing

There's only one part of any presentation guaranteed to capture audience attention – the opening. But most audiences will give you only about one minute to show you have something interesting or important to say. If you don't grab them at the onset, you could lose them for the rest of your remarks.

There are two ways to begin a presentation: traditional and non-traditional.

In the traditional approach, the speaker usually greets the audience, introduces himself, and identifies the topic of the presentation and the agenda or key points he plans to cover. It's part one of the standard, three-part format of a presentation: "Tell them what you're going to tell them, tell them, and tell them what you told them." This is probably the most frequently used opening technique. And there's nothing intrinsically wrong with it (other than its predictability) – especially if it's delivered powerfully.

Then there's the non-traditional opening. In this approach, the presenter begins with something unexpected – for example, maybe a story. The audience is intrigued, and for a few moments, wonders where the speaker is headed. Afterwards, the speaker may introduce herself and identify the topic and purpose of the presentation before continuing. It's unconventional. Don't be afraid to try something different.

Regardless of the approach you choose, don't be long-winded. Keep it short – no more than about ten percent of your total presentation length.

The closing of your presentation should return to the most important idea or ideas covered in the body of the presentation. The reason for this has to do with what's called the "principle of recency." People tend to remember best what they heard last.

That doesn't necessarily mean that you simply "Tell them what you told them." You can repeat your messages, but do so differently, creatively, inspirationally. Find a way to get your audience thinking about what they just heard or learned, or what you want them to do. Don't end with a whimper. End with something memorable.

8. Determine how you'll deliver the presentation

Do you plan to read verbatim from a prepared text (from either a printed script or perhaps a teleprompter)? Or will you be using notes or bullet points as your prompt? Develop the appropriate tool for your delivery.

9. Consider PowerPoint

Okay, now it's time to think about PowerPoint . . . and whether it's needed to help convey your information. If so, can you create your own slides? Not sure? Creating PowerPoint visuals is incredibly easy. This is not a book on how to navigate PowerPoint software in order to create a slideshow. (There are a lot of "how to" books out there that do that.) But a brief tutorial seems in order. You don't need

Those Pesky
Safe Harbor Statements

to be a tech guru or graphic designer to come up with some pretty impressive visuals for your presentations.

On the previous page is the title slide for a presentation about "Safe Harbor" statements. (These are statements publicly traded companies are required to make when they share "forward-looking" information – for example, projections about future earnings.) The slide took about five minutes to create.

Here's how you create a PowerPoint slide:
1. Open the Microsoft PowerPoint app.
2. Click "New" to create a new presentation.
3. Choose either "Blank Presentation" or one of the many templates (themes) shown on the screen. (I chose "Blank Presentation.")
4. Two rectangular-shaped boxes will appear – one for you to type in the title of your presentation and the other for a sub-title if you want one. You can customize the type style and size. Then add your content.
5. If you want to include a photo (as I did), click the "Picture" icon at the top of the screen (the Ribbon) and choose one of the four options listed. (I chose "Picture from file" then went to my desktop and clicked on a photo. It immediately appeared on the slide.) When you do this, PowerPoint will automatically create a theme or look for your presentation and provide you with other theme options (shown on the right of the screen). Any additional slides you create for your presentation will appear in the theme you chose.
6. Select "New Slide" at the top of the screen to continue to build your presentation.
7. When you're done, click on the icon marked "Save this presentation."

And that's it. By walking you through those seven simple steps, I may just have rendered it unnecessary for you to read one or more chapters from most of the 7,000 PowerPoint books listed on Amazon!

Try creating your own slides and you'll see why PowerPoint has become the undisputed leader in presentation software. Once you've created your slides and know how to deliver them effectively, you're good to go.

Avoid Text-heavy Slides

Some well known products began life very differently from how they're currently being used:

- Listerine started out as an antiseptic.
- Bubble Wrap was originally envisioned as wallpaper, and Play-Doh was first used as wallpaper cleaner.
- Frisbees were originally pie containers.
- Fruitcake was a dessert, not a doorstop. (Just kidding on this one.)

Add PowerPoint to that list. Really? Really! PowerPoint was designed for the audience – to help it get and retain information. But it has turned into a prompt for the presenter – loaded with words. Lots of words. This has happened for several reasons: the presenter is not fully knowledgeable about the topic, is unwilling or unable to practice the presentation, or simply observes others relying heavily on what has become the notes on the screen. Talking to the screen, or worse yet, reading to the audience, is usually the result. Let me repeat: that screen and what's on it are for the audience, not for the presenter.

So, are words to be avoided in PowerPoint? Not at all. If you do use words, keep them to a minimum – such as in a news headline or classified ad: No complete sentences. Skip unnecessary words (e.g., "a," "an," "the," etc.). Exceptions could be quotes, a company's mission statement, presentations to non-native language speakers.
Take a look at the following typical slide:

Learning to Sail

- Familiarize yourself with the unusual vocabulary of sailing (e.g., port, starboard, halyard, sheet, jib, close-hauled, etc.).
- Know what each piece of equipment on the boat does.
- Take lessons from a licensed captain or an experienced sailor.
- Venture out when the wind is no more than 15 knots and the waves are two feet or less.
- Practice turning the boat by tacking and jibing.
- Practice docking the boat – both bow in first and stern in first.

It's deadly. For one thing, the audience will start to read it as the presenter is talking. And most people can read at 600 words per minute (WPM), while the speaker is probably talking at 150-200 WPM. So the presenter has essentially lost control of the flow of information. He's talking about sailing vocabulary, but the audience has already begun thinking about sailing equipment.

The human brain can process only one incoming message at a time; an audience can either read your visuals or listen to you. It cannot do both simultaneously.

How to edit that slide:

Go through it and pick out the most important words. In a classified ad, you pay per word, so you find a way to convey your message succinctly. Do likewise when using word slides.

Learning to Sail

- Familiarize yourself with the unusual vocabulary of sailing (e.g., port, starboard, halyard, sheet, jib, close-hauled, etc.).
- Know what each piece of equipment on the boat does.
- Take lessons from a licensed captain or an experienced sailor.
- Venture out when the wind is no more than 15 knots and the waves are two feet or less.
- Practice turning the boat by tacking and jibing.
- Practice docking the boat – both bow in first and stern in first.

Revise the slide:

Learning to Sail

- Unusual vocabulary
- Equipment
- Take lessons
- Venture out
- Practice turning
- Practice docking

Add a photo and now you have a slide you can use to guide your audience through a discussion of sailing. Note that the complete sentences have been eliminated and the number of words has been reduced from 81 to 14.

Another option is using a "build" or "reveal," where you sequentially reveal information on the slide. In other words, each of the six points on the revised slide is added one at a time – allowing you to discuss it without the audience being able to read ahead. This is similar to what presenters did years ago when they used overhead projectors; they used a sheet of paper to cover up portions of the transparency until they were ready to reveal the information. A slightly different build approach: each time you add a new point, the previous points can remain, but appear in lighter type; this helps the audience to focus on your current point.

A better option:

An even better option is to create six individual slides – each with an appropriate photo. Here's one of those slides:

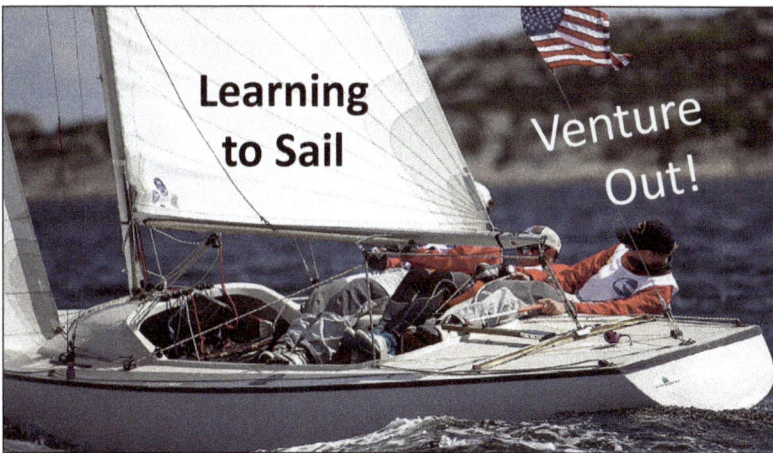

Many presenters cling to those text-heavy slides because they've never realized how ineffective they are. And besides, don't all their colleagues use slides like that? Effective use of PowerPoint requires the presenter to know the material well. A word or two on a slide should be able to trigger fairly detailed comments from the presenter.

Does Your Presentation Need an Agenda Slide?

Probably not. The second slide in nearly all PowerPoint presentations is usually an agenda slide. (The first slide is almost always the title slide.) But most agenda slides are deadly . . . and unnecessary.

There's only one part of any presentation guaranteed to capture audience attention – the opening. But most audiences will give you only about one minute to show you have something interesting or important to say. If you don't grab them early, you could lose them for the rest of your remarks. Don't squander this important part of your presentation on the ordinary or the mundane.

In yesterday's Hollywood, movies began with some opening credits – names of lead actors and behind-the-scenes executives such as the producer and director. There might also be some music along with background information about the story, provided in written form onscreen after those credits – location, time, etc. Not anymore. That approach has been replaced by openings designed to grab you.

In the James Bond movie *Skyfall*, the first thing we see on screen is Agent 007, gun in hand, walking cautiously down a hall. He enters a room, sees two bodies on the floor and a fellow agent bleeding to death. In the next few minutes, Bond chases the culprit – by car, on a motorcycle throughout Istanbul's Grand Bazaar, and on top of a moving train. Eventually Bond is shot (mistakenly, but not fatally, by a colleague) and falls off a railway bridge into rushing water. All of this takes place in the 13 minutes, 11 seconds before we hear Adele sing "Skyfall," and the opening credits appear. Hollywood knows how to get and hold audience attention. We can learn a thing or two from this industry.

Let's say you're delivering a talk about presentation anxiety. You could start with an agenda slide like this one – identifying what you plan to cover:

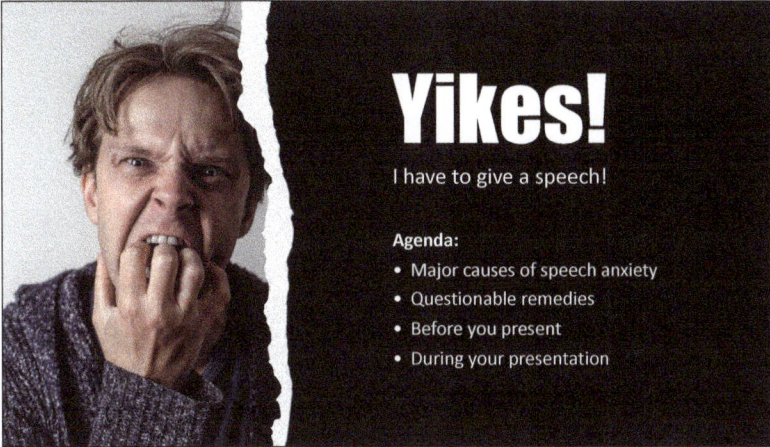

This typical slide (with a cliché photo) is not likely to generate much anticipation. Plus, it's not really needed.

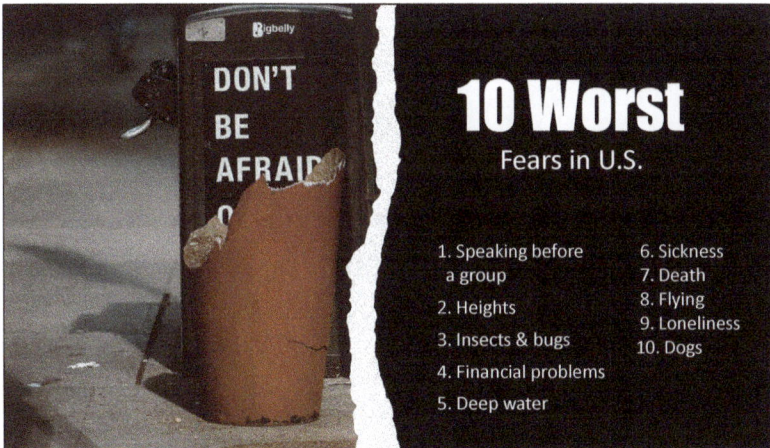

A better way to begin might be to show a list of the ten worst human fears in the U.S. (fear of public speaking tops the list). Grab audience attention by pointing out that the fear of public speaking trumps some pretty serious fears.

Or how about a photo of someone famous who suffered from glossophobia (fear of public speaking)? Thomas Jefferson was so lacking in confidence and afraid of speaking in public that even though he was upset about the changes the Continental Congress made to his carefully written Declaration of Independence, he was unable to speak up and defend his work.

Patient: Jefferson, Thomas

DOB: April 13, 1743

Birthplace: Shadwell, Virginia

Height: 6'2"

Spouse: Martha

Occupation: Lawyer, Architect, Founding Father

Diagnosis: Glossophobia

After commenting on either of those images, simply tell your audience what you plan to discuss. No agenda slide is needed for those four bullet points. You can even repeat each of those points at the appropriate time during your presentation by saying something like, "Now let's take a look at the major causes of presentation anxiety."

One year to the day of George Harrison's death, the memorial *Concert for George* took place at the Royal Albert Hall in London. Eric Clapton, who served as master of ceremonies, welcomed the audience and provided a preview of the program – a vocal agenda. He said, "First of all, there will be an Indian section. The first half will be some music that was composed especially for this by Ravi Shankar. His daughter Anoushka is going to play. Then there will be a little intermission, and then we'll have some Western music. We'll be playing George's songs and we'll be having some guests. I'll let you spot them, and I'll introduce them as they come on."

That opening was powerful in its simplicity. And it was spoken – not read – without the use of notes or PowerPoint visuals. The opening of your presentation can be just as powerful – without an agenda slide. Command the attention of your audience with some powerful words or images delivered powerfully.

Longer presentations, workshops and seminars might lend themselves to an agenda slide (or a printed agenda). But if you go that route, don't lead with your agenda. Come up with a powerful opening, then transition to the agenda.

Dodging Bullets

Two bullets about bullet points:

- Nearly every PowerPoint presentation has them.
- PowerPoint gurus disagree about their value.

Let me weigh in on this.

The term "bullet point" refers to the dot placed in front of a word, phrase or sentence. These dots resemble bullets being shot from a gun. In documents, bullet points serve a variety of purposes. They may signal a sequence or a summary. Or they can be used to break up long, uninviting text material – making it easier to process, understand and remember. Bullet points capture reader attention. You're probably using them in your resume to highlight important details.

Because bullet points are an efficient way of communicating – short, to the point, memorable – they have migrated into PowerPoint presentations. Here are some not-so-hard-and-fast rules for using bullet points in a presentation (provided in bullet point format, of course):

- Avoid bullet point-use default – automatically using bullet points on all or most of your visuals.
- Use them with intent – for example, to show items in a list, to summarize something, or to highlight important points.
- Use numbers rather than bullets if you're counting something or if sequence is important.
- Use parallel structure. If the first bullet is a sentence (or a phrase or word), the subsequent bullets should be in that format as well. Also, each bullet should begin with the same part of speech (i.e., noun, verb, etc.).
- Capitalize the first letter of each line. Subsequent words in the bullet point (except for proper nouns) should be lower case.
- Reserve periods (punctuation) for complete sentences.
- Avoid sub-bullets.

Avoiding Animation Overuse

Most of us are fascinated by animation. Who doesn't fondly remember a character or scene they watched in a Disney movie such as *Pinocchio, Bambi, Cinderella* or *The Lion King*? So, it's no wonder that animation, which uses movement to convey an idea, has become a popular element in PowerPoint presentations.

For example, if you want to show a photo of your new corporate headquarters, why not first show the audience a picture of the old headquarters? Then, have that picture slowly fade to black, and then fade back in with a photo of your new digs. The message: this is a successful company and it's on the move. PowerPoint enables you to communicate change and other concepts through motion on a slide or when you move from one slide to another.

When Al Gore ran for president, most Americans noticed his seriously sub-par communication skills. On the campaign trail, he came across as stiff, professorial, unrelatable. After his defeat, he returned to the campaign trail – but this time campaigning to increase awareness of climate change. Amazingly, he became one of the world's most effective communicators on the issue of global warming.

He did it by delivering a climate change slide show to hundreds of audiences worldwide. That slide show eventually became a feature of *An Inconvenient Truth*, an Academy Award-winning documentary. The film has high production values – innovative staging, powerful music and first-rate visuals, including animation - <u>appropriate</u>, <u>effective</u> animation.

Gore wanted to point out that a lot of people wrongly assume that the Earth is so big, humans can't possibly have any lasting impact on its environment. So he told this story: One of his grade school teachers taught geography by pulling a map of the world down in front of the blackboard. A classmate raised his hand, pointed to the outline of the east coast of South America and the west coast of Africa and asked, "Did they ever fit together?" The teacher said, "Of course not; that's the most ridiculous thing I've ever heard."

Gore went on to explain that the teacher was actually reflecting the conclusion of the scientific establishment of that time – namely, that continents are so big, obviously they don't move. Then he added, "But actually, as we now know, they did move; they moved apart from one another, but at one time they did fit together." As he said that, animated images of South America and Africa were projected on the screen – moving toward each other and eventually fitting together perfectly.

Like Gore, I found an opportunity to harness the power of animation.

You've probably heard of or watched a TED Talk (www.ted.com). My company's version of these amazing presentations is the KEN Talk. One of the talks I deliver is titled, "Presentations That Rock: What Business Leaders Can Learn from Rock Concerts."

It examines nine elements of successful rock concerts, their parallels in business presentations, and how business professionals can use some of the strategies and tactics of rock musicians to deliver a dazzling presentation. To drive home those parallels, I use animation. Each of the nine rock concert elements is introduced with a term that moves across the screen and morphs into its equivalent term in a business presentation. So, for example, "Set List" (the songs featured in a concert) becomes "Content." "Intermission" becomes "Break." "Exit Song" becomes "Conclusion." "Encore" becomes "Q&A." Etc.

The key to using animation correctly is to make sure it helps the audience process and understand the information being presented. Having words, phrases or images move around or zoom in or out on a slide simply because the technology exists to make that happen is a gimmick, not a strategy for enhancing learning. Human beings are hard-

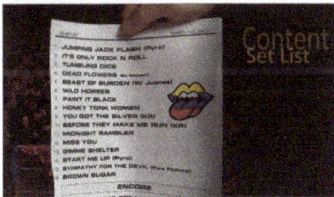

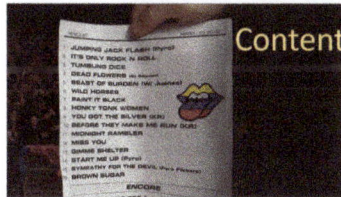

wired to look at things that move; make sure your audience is not distracted by motion, but instead is helped by it. Animation can help you communicate a variety of things: passage of time, change in direction, etc.

San Francisco 49ers quarterback Brock Purdy at a post-game press conference – competing with multiple logos for audience attention.

Rein in That Logo

Watch any NFL team's pre- or post-game press conference and you'll see a coach or player standing in front of a background containing repeated images of the team's logo and logos of one or more sponsors. (In football, as in business, marketing is everything.) But these logo-laden backgrounds could be counterproductive. Human attention is limited; given the incredible amount of information we're bombarded with, our brains have evolved not to help us take more information in, but to help us screen some of it out. Providing more information, when human attention is already in short supply, might hamper audience ability to receive, process and remember the speaker's message.

In most PowerPoint presentations, particularly those developed from a company template, attempts at "subtle" marketing often take the form of a logo on every slide. Try to avoid this mistake. Here's why:

With internal audiences, your colleagues certainly know what organization you represent. And external audiences probably know as well – from your introduction, your opening slide or the printed or digital session agenda. So why repeatedly remind them where you work?

Also, a logo on every slide tends to commercialize your presentation – regardless of its mission or content. Think about how you react when you attend some professional meeting or event (especially one you paid for) and a program speaker delivers a presentation that's nothing more than a thinly veiled commercial for his company. You feel a bit cheated. Overusing your logo can negatively impact how the audience perceives you or your message.

A fundamental goal of any communication is to get and hold audience attention. It's harder to do that today than it was in the past. All those superimposed TV channel logos and "crawls" (moving messages) that appear on screen during your favorite program? They're competing with on-screen content for audience attention. Ditto for that repeating logo on every slide, and for those barely readable footnotes found on the bottom of some slides. (If possible, put the additional detail in those footnotes on the audience handouts, but keep it off your projected images.)

Avoid cognitive overload (and PowerPoint slide clutter). Put your logo only on your first and last slides.

Your Final Slide

The principle of recency, one of several principles of learning, says that people remember best the last thing you tell them. That suggests the ending of your presentation is critically important. But most presenters give it short shrift. Big mistake. Part of the blame for weak endings frequently has to do with the visual a presenter uses – specifically, the last slide.

Here are some typical presentation endings, along with a critique:

1. **Abrupt, Confusing Ending.** Some presentations just "die." At the end, the presenter stops (usually awkwardly), and the audience is left wondering, "Is that it? Is there anything else? Is it over?" Former U.S. Treasury Secretary Timothy Geithner

delivered a critical speech detailing the plan the federal government developed to address the economic crisis that was crippling the U.S. and world economy in 2008-2009. At the conclusion, he stopped abruptly, turned and walked away. The puzzled audience eventually figured out it was over and that they could leave. Don't keep your audience wondering. Make a clear and obvious change in gear. You can do that in a number of ways — including a change in content (e.g., summarizing key points, providing a powerful story, etc.), a change in delivery tone (e.g., speaking rate, volume, pitch), or some inspiring words or a call to action. The best endings are those where the audience can tell that the presentation is winding down.

2. **"That's all."** Those might be the two words most frequently uttered by speakers at the end of their presentations. And the words are sometimes accompanied by a slide that reads, "The End." While Hollywood still likes to see "The End" projected on screen at the conclusion of a movie (before the final credits roll), those words (whether uttered or written) have no place in a presentation. Avoid them.

3. **"Thank You."** After a speech or presentation, it's just good manners to thank the audience for their time and attention. But skip the "Thank You" slide.

4. **Q&A.** Q&A is an important element of most presentations — in some cases, the most important element. It's two-way communication. It provides your audience with an opportunity to participate, and you with an opportunity to get feedback. But don't keep the slide containing your concluding content up on screen during the

Q&A period. The audience will keep looking at it occasionally. Instead, black the screen. Also, there's no need for a slide that says, "Questions?" or "Q&A." Simply say you'll be happy to field some questions. And if you delivered your presentation from behind a lectern, step out in front of it and get closer to the audience. That gesture speaks volumes and can motivate the audience to ask questions.

5. **Summary.** Many speakers end their presentations with a slide that recaps their core message or key points. For example, if you delivered a presentation on work/life balance, your final slide might contain a list of the advice you discussed along with an imaginative photo:

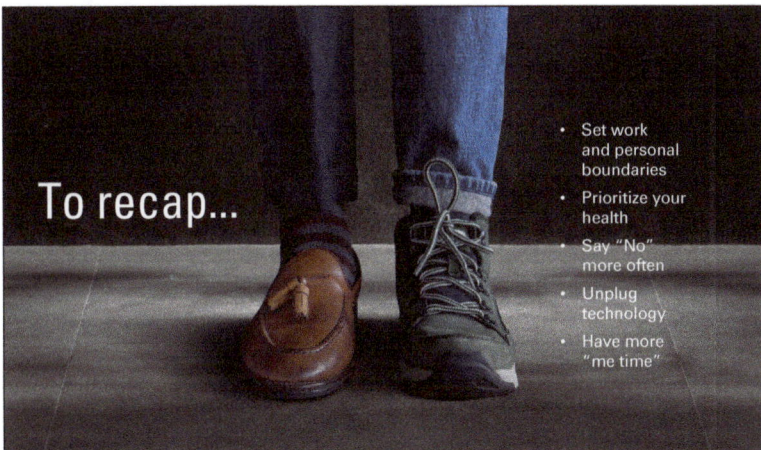

To recap...

- Set work and personal boundaries
- Prioritize your health
- Say "No" more often
- Unplug technology
- Have more "me time"

This approach makes sense; it's the third part of that widely used presentation format that says, "Tell 'em what you're going to tell 'em. Tell 'em. Tell 'em what you told 'em." Repetition is reinforcement.

6. Creative Ending. But how about a more creative ending than part three of that "Tell 'em³" format? In this approach, rather than have a summary slide, remind the audience of your key points, then share some inspirational comments, perhaps along with an on-screen quote. Here are two final-slide candidates for that work/life balance presentation:

"Never get so busy making a living that you forget to make a life." –Dolly Parton

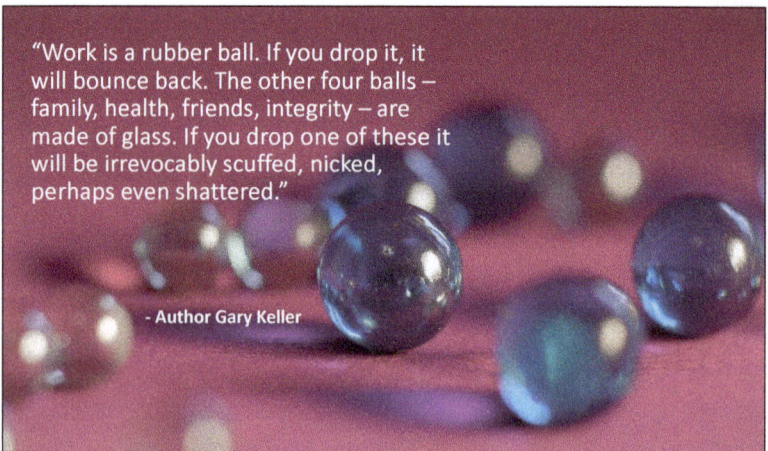

"Work is a rubber ball. If you drop it, it will bounce back. The other four balls — family, health, friends, integrity — are made of glass. If you drop one of these it will be irrevocably scuffed, nicked, perhaps even shattered."

- Author Gary Keller

The marital breakup of superstar quarterback Tom Brady and Gisele Bündchen was front-page news. It was no secret that for several years, Giselle was concerned about how Tom's commitment to football was negatively impacting their family life. Things only got worse when Tom "unretired" from football after just a few months of retirement. The final slide of my hypothetical work/life balance presentation might have been a photo of the Brady family during happier times, accompanied by some carefully chosen, compassionate comments about what happened and what can be learned from it. Sometimes, current events can serve as "sparklers" for your presentation. Look around for them.

7. **Logo.** A few pages earlier, I recommended that you place your company logo only on your first and last slides. But do you always need to end with a logo slide? For internal audiences, probably not. For external audiences? Perhaps, depending on the audience and the nature of your presentation. For example, sales presentations might benefit from one last promo for your company, its website or some other important piece of information. In short, be selective when it comes to a final logo slide.

A Pictorial "Thank You"

Remember that KEN Talk on "Presentations That Rock: What Business Leaders Can Learn from Rock Concerts" mentioned a few pages earlier?

   While developing that talk, I wondered how it should conclude. Ending with a slide that summarized the nine elements of a rock concert and their equivalents in business presentations just wouldn't fit with the highly entertaining content of the talk and its performance-style delivery (including using video clips from concerts). I needed a different kind of final slide – one that required no projected words or accompanying commentary. Here's the photo that appeared on that slide. To paraphrase a well known saying about the value of a picture, "This picture is worth . . . two words: Thank You." Think creatively when it comes to your final slide.

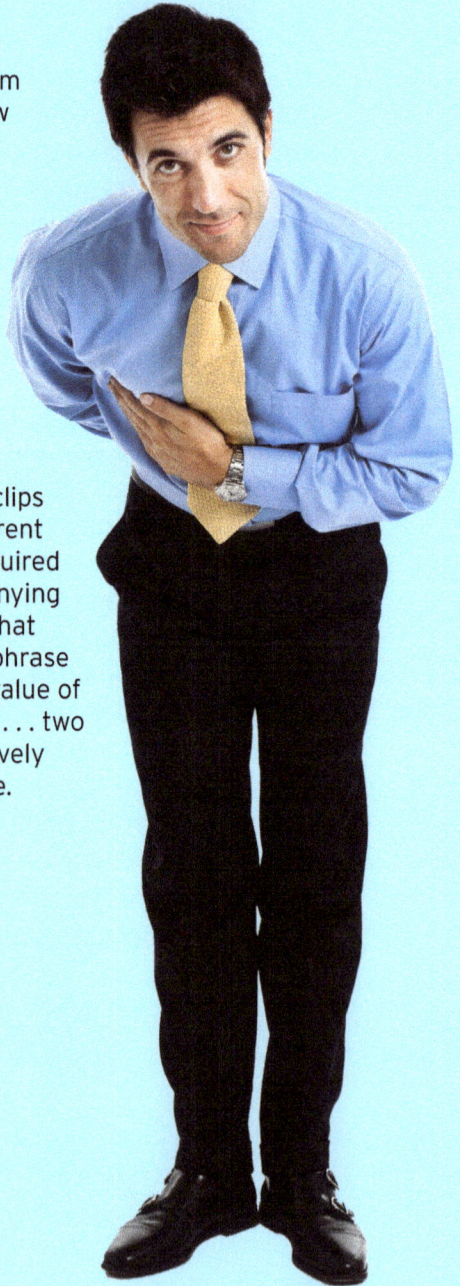

OH NO, NOT POWERPOINT! RELAX, HELP IS ON THE WAY.

Congrats!

You're halfway through the book.

Are its ideas resonating with you? Are you ready to make some changes in how you use PowerPoint?

Four chapters to go. Let's explore a bit more, including how to create more striking slides and deliver them more effectively.

The brain processes visuals 60,000 times faster than it processes text.

Chapter 5

Create Visual Impact

Images vs. Text: And the Winner Is . . .

You've probably noticed that most PowerPoint slides, including perhaps the ones you use, contain lots of text. Text is made up of small shapes called letters, which are combined to form larger shapes called words. Words, either single or grouped together, impart meaning; they may represent something physical (such as a "chair") or something abstract (such as "justice").

When you read a piece of text, your brain performs an incredibly complex decoding process. For example, if the word, "chair" appears on a slide, your brain must first process the shape of each of the five letters as well as their arrangement into a word. Next, your brain must determine the meaning of that word. But some words have multiple meanings ("chair" can be a noun or a verb), so additional processing is necessary. And if English is not your native language, you likely have to identify the equivalent word for "chair" in your first language. All of this happens very quickly, but there's an even faster, more efficient, more effective way to help an audience derive meaning: show an image of a chair. The brain processes visuals 60,000 times faster than it processes text.

Speaking of chairs, right before the first anniversary of Russia's invasion of Ukraine, Ukraine President Volodymyr Zelensky went to London and addressed Britain's Parliament – seeking continued support for his country's fight to remain independent. In his remarks, Zelensky told lawmakers how on a previous visit to the Churchill War Rooms, a tour guide invited him to sit down in a chair used by Winston Churchill during World War II. Zelensky obliged, and when he was asked how he felt, he shared some thoughts about bravery, hardship and victory.

If you were delivering a PowerPoint presentation about Zelensky's communication skills as a wartime leader (and they are considerable), which of the following two slides would you choose while sharing that powerful story?

Volodymyr Zelensky

- Fluent in Ukrainian, Russian, English
- Earned law degree
- Worked as screenwriter, actor, comedian, director
- Elected Ukraine president (2019)
- Compared by some to Churchill

From **Comic...**
to **President...**
to **World-stage Communicator**

If you chose the photo, good choice. People (and their brains) prefer images over text.

Stock Photos: The Good ... the Bad ... and the Illegal

PowerPoint excels at delivering information that lends itself to visual representation – pie charts, bar graphs, flow diagrams, illustrations, organization charts, and of course, photos. Photos are especially popular with PowerPoint presenters, so let's talk about the good, the bad, and the illegal associated with using stock images.

Stock photos are simply existing photographs that are archived, usually by stock photo agencies. (A few of them are listed below.)

Here's how stock photography works: Photographers, who own the copyright to their work, provide agencies with their photos, and are compensated when someone purchases a license to use the photo. Many stock photos are royalty-free, which means that you don't pay a royalty to the photographer; you simply pay the licensing fee to the agency. Licensing fees vary, depending on the image and how it will be used. Some commercial uses, such as putting the photo on product packaging or on a book cover, are usually prohibited.

Using a stock photograph in a presentation might run about $20. (Other uses, such as for websites, magazines or books, and marketing typically range from about $50-$200.)

As you can see, purchasing the license to use a photo for a PowerPoint presentation is relatively inexpensive. Plus, obtaining your images from an agency allows you to use the image legally. (Most stock photo companies put a watermark across the images that appear on their website; the watermark disappears when you pay for and download the image.)

Stock photos can also be accessed and downloaded via Google and other search engines. The image may or may not be accompanied by a reminder that it might be copyright protected. That means if you use the image, you are responsible for finding and compensating the photographer – not an easy task.

Without a doubt, many photos used for PowerPoint presentations are being downloaded from Google and are probably being used illegally. The user either is unaware of copyright infringement issues or feels that it's unlikely the use will be discovered. The latter may be true with PowerPoint presentations that don't go beyond inside-company use. But what if the presentation makes its way onto the company's website or an audience member captures the image and posts it on social media? It's probably worth mentioning that "ambulance chasers" (i.e., lawyers) are out there looking for copyright infringement violators (although probably not PowerPoint users); it's becoming a lucrative specialty practice. In short, be careful if you download and use an image directly from Google or another search engine.

Fee-based Stock Photo Sites

Alamy (www.alamy.com) British stock photo agency launched in 1999. Provides rights-managed and royalty-free stock photos, 360-degree images, vectors (images that don't lose quality when resized) and videos. Claims to add 100,000 new images daily from suppliers in 173 countries. Customer service includes experts who provide help via email, phone or live chat.

Getty Images (www.gettyimages.com) Established, well known company that supplies stock images, editorial photography, video and music. Has the world's largest privately owned archive of historical photos. The company targets creative professionals, the media and corporate clients. Getty is one of the leaders in the stock photo industry and its higher prices reflect that position.

Shutterstock (www.shutterstock.com) One of the most popular and affordable stock photo and music agencies on the internet. Originally a subscription site only, the company now offers a la carte pricing. Site is in 21 languages.

Dreamstime (www.dreamstime.com) Provides royalty-free photos, illustrations and vectors "at prices that anyone – from large corporations to various magazines and blogs – can easily afford." Has a free images section that provides access to one of the few collections of free images.

Adobe Stock (www.stock.adobe.com) Stock photo service offered by software company Adobe. Requires monthly or annual subscription.

Fotosearch (www.fotosearch.com) Provides royalty-free and rights-managed stock photography, illustrations, maps, video and audio from publishers located around the world. The company offers both single-image and subscription pricing.

iStock (www.istockphoto.com) Online, royalty-free provider of photos, illustrations, clip art, videos and audio tracks. Acquired by Getty Images in 2006, and headquartered in Calgary, Alberta, Canada.

Associated Press (www.ap.org) An independent global news organization, AP operates a photo division with a collection of more than 35 million editorial, sports, entertainment and creative stock photos. Its archive contains some of history's most iconic images.

"Free" Stock Photo Sites

Some photos from the following companies are completely free. However, the number of free images tends to be limited. During your search, you'll also be shown premium photos (some from partner sites) that require a fee. Once you specify a particular image you're looking for, these sites are likely using AI to place the best photos in the premium category.

Unsplash (www.unsplash.com) Owned by Getty Images, Unsplash is a pioneer of the copyright-free photography model. Downloaders can "copy, modify, distribute and use photos for free, including for commercial purposes, without seeking permission from or providing attribution to the photographer or Unsplash." Impressive collection of photos of nature, landscapes, cityscapes and scenery.

Pixabay (www.pixabay.com) Website that provides free images, illustrations, videos and music. All free content can be used safely without asking for permission or giving credit to the artist — whether for commercial or non-commercial purposes.

PikWizard (www.pikwizard.com) Provides high-quality stock photos, royalty-free and safe for commercial use, with no attribution required. Full access to the site's image-editing tool requires a fee.

Pexels (www.pexels.com) High-quality, completely free stock photos that come from the site's users or are sourced from free-image websites. Pexels' database is also available in the Canva app.

Gratisography (www.gratisography.com) Free, stock photo website that offers "the quirkiest, weirdest, and funniest" stock photos on the internet. Small, but unique collection.

Planning to Use Photography?

Inserting a photo onto a slide is easy. PowerPoint lets you do it in a few clicks. But selecting the right photo – that's a challenge. Some tips:

- Don't treat photos as an afterthought – to fill white space on a slide. They are not decoration. Use them for a specific purpose, such as when a photo rather than words can do a better job of conveying an idea. For example, if your company encourages its employees to volunteer in their community, the best way to convey that idea might be through a photo of actual employees at a Habitat for Humanity building site.
- When possible, choose realistic photos over generic images. Many stock business photos show "beautiful people" in staged, unrealistic situations.
- Some companies hire professional photographers to get images for their website, annual report and other internal and external communication tools. Check to see if your company has a library of photos you can use in your PowerPoint presentations. These libraries (typically managed by the advertising, PR or marketing function) generally contain pre-approved and pre-paid photos of actual people, facilities and events in your company – a real plus.
- Choose images that are contemporary and reflect today's more diverse culture and ethnicity.
- Avoid predictable, overused photos.

Cliché. It's a word borrowed from French, originating in the printing trades. It refers to a plate or block that repeatedly reproduces type or images. We use the word to mean something that's overused to the point of being irritating and that suggests a lack of original thought.

"Think outside the box." "Win-win situation." "Not in my wheelhouse." Those are a few business clichés that have outlived their usefulness. Some photos typically used in PowerPoint presentations also have outlived their usefulness:

the light bulb

the handshake

the globe

the bullseye

smiling workers

Let's say you're looking for an image that communicates: ideas, creativity, thinking. Why not replace that dim, worn-out light bulb with something like the following?

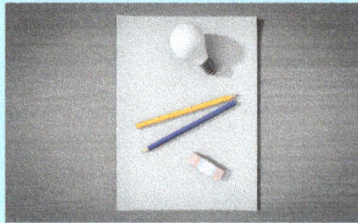

Any ideas on alternates for those other four cliché photos?

- Choose photos that grab attention and reinforce the message you're trying to convey. Book cover images are carefully selected (to sell books). Do likewise with every image you use.
- Be careful of those cheesy, popular montages – multiple photos on a single slide. They rarely work, in part because they're overused and are hard to see and process.
- Make sure all your photos have a high-quality (e.g., high resolution), professional look.

Charts & Graphs: A Brief Tutorial

The good news: The tools available to communicate data visually are many and varied: pie charts, bar graphs, maps, flow diagrams, org charts, scatter plots – to name a few. The bad news: Most of these dashboard tools are not being used effectively. That's not surprising; most of us have never been shown when to choose and how to design these tools of business intelligence.

What follows is a brief tutorial on how to display data in some of the most frequently used graphic slides. Warning: After reading this, you may want to do some chart makeovers for your next PowerPoint presentation.

Pie Charts

Pie charts are the workhorses of the chart world. They're one of the most widely used data visualizations. They're also the one most widely criticized by experts who say pie charts communicate information poorly.

A pie chart is essentially a circle divided into segments, where each segment represents one component and all segments added together equal the whole. Pie charts show how large or small something is compared to the whole – but they do it generally rather than precisely. In other words, pie charts are limited to communicating whether something is large, small or about the same. More exact comparisons are difficult to discern. The following pie chart works well for communicating the components of credibility. You get a good feel for the relative weight of each component.

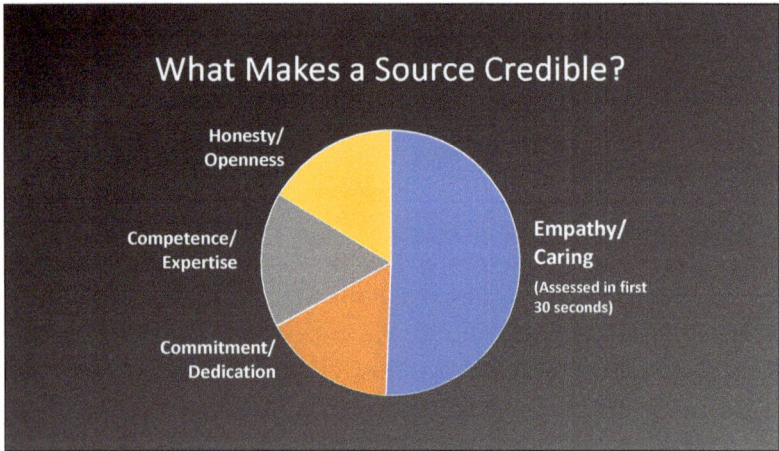

One of the limitations inherent in pie charts has to do with the fact that the human brain can compare length easier than it can compare angles. Look at the pie chart below on the left.

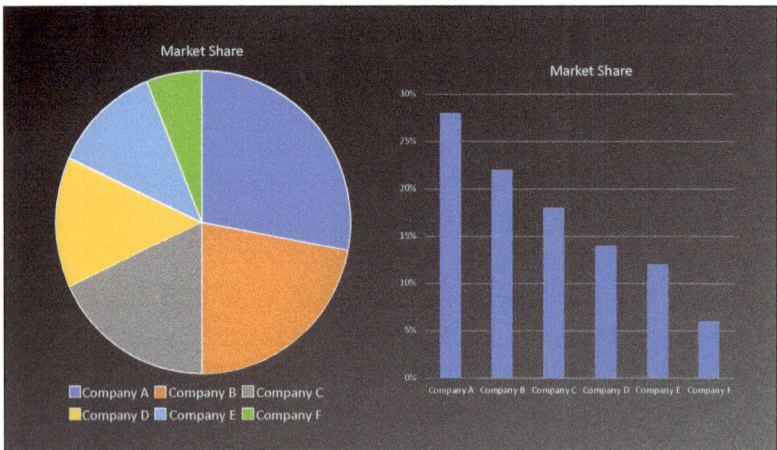

Most people would conclude that some of the segments are pretty close in size. But if we take that same data and convert it into the bar graph on the right, comparison becomes easier and we can see some significant differences.

Note to corporate managers responsible for developing visuals for their company's financial presentations: If your sales and revenues or earnings are greater than those of your competitors – but only marginally greater – pick the right tool. Opt for the bar graph over the pie chart.

Best Practices

- Limit the number of pie segments. A maximum of eight is a good rule of thumb. Anything more could be difficult to read.
- Organize your data from largest to smallest. Begin your first data segment at the top middle and continue clockwise.
- Use caution with 3-D effects. "Tilted" pie charts tend to distort (i.e., emphasize) data in the foreground. With "exploded" pie charts, one or more sections are separated from the whole. This will highlight a section but make it difficult to judge the part-to-whole comparison.
- Try to avoid using a legend when labeling pie segments. Label the segments directly so the viewer doesn't have to look back and forth between the chart and the legend to discern the meaning (as in the Market Share pie chart on the previous page).
- Choose colors carefully. Avoid using the same or similar colors within the chart. Also, stay away from bold primary colors.
- Don't compare two pie charts side by side. The charts you're trying to compare are separated and the segments are probably located in different parts on the two pies.
- Make sure the portions add up to 100 percent.

Bar Graphs

A bar graph uses bars or lines of a certain length to compare various categories of data using some measured value.

Bar graphs do a better job than do pie charts of comparing the size of categories (i.e., whether bar A is larger or smaller than bar B). They also do a better job of comparing data across data sets. (For example, comparing slice A from one pie with slice A from another pie would be difficult.)

Bars can be displayed either vertically or horizontally. One advantage of horizontal bar graphs is that you can avoid category labels that are hard to read (i.e., vertical or angled) or that result in unusually wide graphs. Here's a horizontal bar graph that displays purchasing data in a way that's easy to interpret:

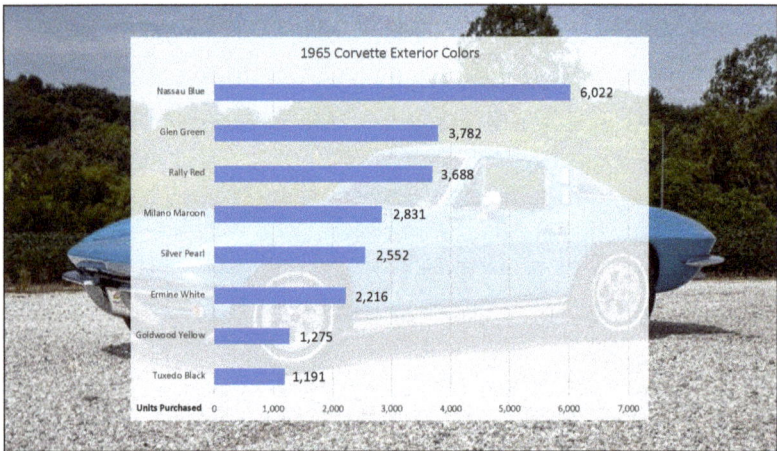

1965 Corvette Exterior Colors

Color	Units Purchased
Nassau Blue	6,022
Glen Green	3,782
Rally Red	3,688
Milano Maroon	2,831
Silver Pearl	2,552
Ermine White	2,216
Goldwood Yellow	1,275
Tuxedo Black	1,191

Best Practices

- All the bars on the graph must be plotted on a zero-value baseline.
- The top or end of each bar should be flat (not 3-D, rounded or some other shape) so the viewer can easily determine its value and compare bars.

- When assembling a bar graph, organize the bars from longest to shortest.
- Generally, the bars do not need to be different colors. Length provides the contrast among bars.

Organization Charts

One chart that usually makes its way into many company overview presentations is the organization or org chart. It's a visual that shows how a company is structured – how functions or departments are organized and which individuals run them.

The different types of org charts include:

- Hierarchical: the most common, showing the highest-ranking individual at the top with subordinates below.
- Horizontal or flat: depicts an organization with little or no middle management.
- Matrix: combines hierarchical and horizontal, showing top-down reporting relationships and cross-functional teams headed by project managers.

Organization charts projected on screen rarely communicate information effectively. The problem is that these charts are usually incredibly complex – containing a lot of information crammed into small boxes, sometimes with head-shot photos. The charts are sterile and hard to read. Presenters often gloss over them and quickly move on to the next slide. This is one tool that needs a serious makeover – both in design and delivery.

Best Practices

- Highly detailed and complex org charts work best as leave-behinds where the information can be carefully reviewed. On-screen versions should be treated a bit differently.

- Break complex charts into manageable parts. For example, the following org chart shows the structure of a typical crisis management team, but does so in two parts. The first slide shows permanent positions – the ones that should always be represented when the team meets. In a briefing on crisis management, a presenter could discuss each of those positions in detail while the slide appears on screen.

Crisis Management Team

Team Leader — Assistant Team Leader

Crisis Center Head — Incident Secretary

Legal — Communications

The second slide shows the additional functions that might join a team depending on the crisis. And again, a presenter could elaborate on those functions.

- Most personnel org charts provide names, titles and possibly employee photos. If a chart shows just a few individuals (or if you segment a large group as mentioned above) why not include a brief mention (vocal) of some accomplishment or something else notable for each individual? Find a way to make the org chart interesting and more valuable. Remember, an individual's

Crisis Management Team

Team Leader — Assistant Team Leader

Crisis Center Head — Incident Secretary — Functional Reps (ad hoc)

Legal — Communications

Functional Reps (ad hoc):
- Executive Office
- Human Resources
- Investor Relations
- Sales/Marketing
- Manufacturing/Operations
- Engineering
- Research & Development
- Finance/Accounting
- Health, Environment, Safety
- Security
- Government Affairs
- Information Technology

accomplishments are more interesting and memorable than his or her responsibilities.

- Be selective. It may not be necessary to reference every person or function on the chart. Consider the audience and highlight what's appropriate.
- When most people think of org charts, they picture the standard pyramid format with multiple levels of data. But today a wide variety of templates are available to create charts that are visually appealing. Find those templates on the internet.

Maps

If your goal is to convey geographic information, maps are usually the tool to choose. They are simple, compelling and easily recognizable.

Best Practices

- Choose different symbols and easily distinguishable colors for data point markers.
- Don't overlap data points.
- Avoid too many data points on a small map. Sometimes a series of maps – grouping data by continent, country, region, etc. – is preferable to a single map.
- Maps can serve as backdrops – allowing you to link photos and other types of charts and tables to a specific location.

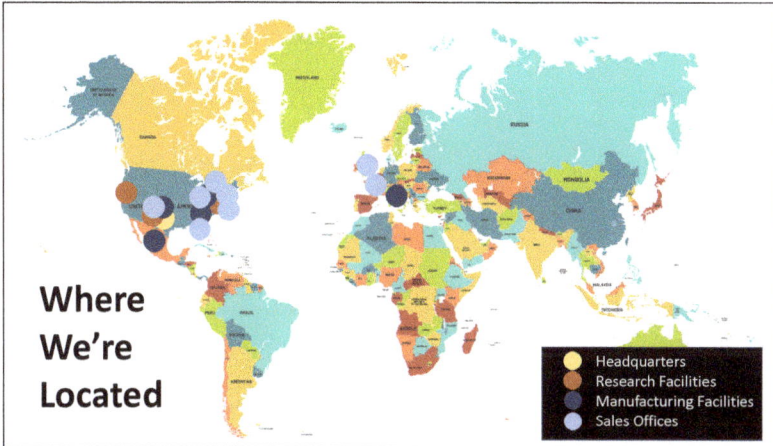

Where We're Located

This map does a poor job identifying where the company's facilities are located:

- A map of the entire world is not needed.
- All facilities are represented by the same type of symbol (dot) and symbol colors are difficult to distinguish.
- Knowing a facility's exact city, state or country is not always possible.
- Some data points overlap.

Our Locations

⭐ Headquarters
   Dallas
● Research Facilities
   San Francisco
   Austin
   Cambridge, MA
▣ Manufacturing Facilities
   Houston
   Cleveland
   Atlanta
   Monterey, Mexico
   Milan
△ Sales Offices
   Chicago
   Houston
   Miami
   Toronto
   Boston
   New York
   London
   Paris

These revised map close-ups do a better job providing important detail:

- The maps focus only on North America and Europe, where the company has operations.
- Facilities are represented by different symbols, are grouped by function, and cities where they are located are named.

Tables

"A necessary evil." That phrase, meaning something unpleasant but required, aptly describes many of the tables that appear in PowerPoint presentations. These tables typically contain columns of numbers – lots of them – and while the data can be important, even critical, it can overwhelm and put off the viewer.

Unlike many charts and graphs, which communicate information visually at a glance, tables deal with details – exact numbers or precise estimates, which sometimes reveal trends or anomalies. Tables work best in written documents, but if carefully crafted, they can be effective PowerPoint visuals.

Best Practices

- Tables that are projected should not contain too much data. Ask yourself whether every element (e.g., column or category) or degree of precision is really needed. (For example, can sales be shown quarterly rather than monthly? Can you round off the numbers?) When assembling data, know what overarching message the table is there to convey. Less is usually more. Also, avoid clutter – such as repetitive percentage and dollar sign symbols. Multiple tables may be needed instead of one complicated table.
- When the table appears on screen, state clearly and concisely what its purpose is, and identify the key takeaway. Don't expect the audience to infer it. And don't feel the need to discuss every element in the table. Focus on what's critical.
- Guide the viewer through the table. Highlight specific data points in bold type or color, or use a laser pointer to direct the viewer's eye.
- Most presenters fail to give viewers enough time to process tables. It takes time for the audience to

make sense of what you're showing them, so don't move on to your next slide too quickly. Be sure to consider this when establishing the length of your entire presentation.
- Make sure the tables you project are included in the leave-behinds.

Income Statement

Years ended December 31, 2020-2023	2023	2022	2021	2020
Net sales	$765,000	$725,000	$602,000	$584,000
Cost of sales	535,000	517,000	421,400	400,800
Gross profit	230,000	208,000	180,600	183,200
Operating expenses				
Depreciation & amortization	28,000	25,000	27,000	27,000
Selling, general & administrative expenses	96,804	109,500	140,000	145,000
Operating income	105,196	73,500	13,600	11,200
Other income (expense)				
Dividends & interest income	5,250	9,500	7,500	8,000
Interest expense	(16,250)	(16,250)	(17,000)	(17,000)
Income before taxes	94,196	66,750	4,100	2,200
Income taxes	41,446	26,250	11,000	9,500
Net income	$52,750	$40,500	($6,900)	($7,300)

This PowerPoint table works: Amount of data is not excessive. The company's most recent results are highlighted in bold. Color is used to flag sales and income loss impacted by COVID.

Slide Makeovers

If you're a frequent PowerPoint viewer, you've probably encountered slides that were a bit "off." Maybe you weren't sure why or didn't even notice, but they likely kept the presenter from communicating effectively. Design problems with slides are many and varied. Here are some common mistakes and how to correct them:

Before: A typical title slide for a company profile presentation. This one has a wrong-sized logo that dwarfs the slide's "vanilla" title (True West: Energy Industry Leader) which doesn't contrast well with its background. Stock photo is inappropriate since the company has no offshore operations.

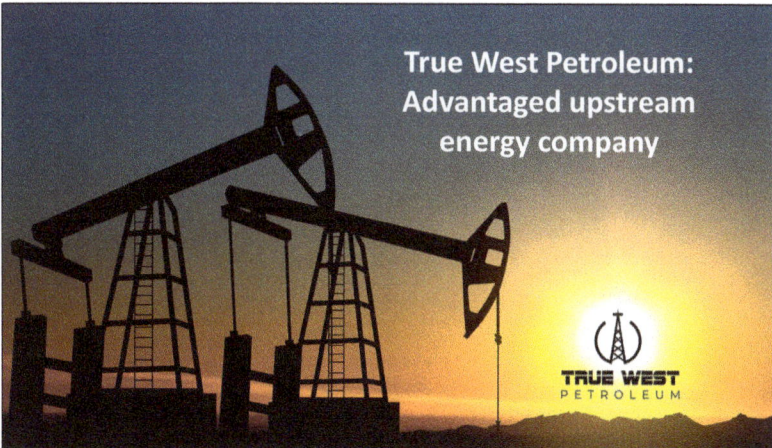

After: Note the slide's revised title. (Which title — original or revised — is likely to pique audience interest in the company?) Text is more readable. Also, the logo has been right-sized and repositioned.

Company Overview

True West Petroleum is an Energy Industry leader committed to sustainable, long-term growth.

Focus: Our strategic focus is on U.S. natural gas – a plentiful, low-volatile, exportable energy source now and for the long term.

Operational Excellence: We continuously improve the reliability of our operations through predictive failure analysis and preventive maintenance.

Safety: The safety of our employees, assets and the communities where we operate is our highest priority.

Growth: We take a disciplined approach to Capital Expenditures, while always looking for appropriate growth opportunities.

Financial: We believe in growing shareholder distributions and consider share repurchases after capital, dividend and debt reduction objectives are met.

Before: Much of the content of this company overview consists of platitudes. (Is there a company out there that isn't "committed to sustainable, long-term growth"?) Focus is on beliefs rather than accomplishments. Inconsistent typeface and text capitalization. Complete sentences and logo add clutter.

True West Petroleum at a Glance

Who We Are: Independent oil and gas E&P company

Our Focus: North American natural gas: globally cost-advantaged fuel

Our Advantage:
Operating in 3 of the most competitive U.S. resource plays: Eagle Ford, Bakken, Permian

Operations:
- Proved reserves continue to outpace production
- Record Permian well production (2023)
- Record Bakken lateral length (2023)

Financial:
- Performing at top of E&P peer group and S&P 500
- Retired >$4 billion of debt (2023)

After: Revised content better showcases the company's strengths. Unnecessary words, capitalization and typeface inconsistencies, logo and cheesy clip art have been eliminated.

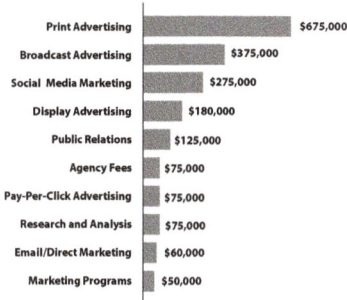

BUDGET SUMMARY

Print Advertising	$675,000
Broadcast Advertising	$375,000
Social Media Marketing	$275,000
Display Advertising	$180,000
Public Relations	$125,000
Agency Fees	$75,000
Pay-Per-Click Advertising	$75,000
Research and Analysis	$75,000
Email/Direct Marketing	$60,000
Marketing Programs	$50,000

Category	Total Spend	% of Budget
Research and Analysis	100,000	4%
Print Advertising	650,000	34%
Broadcast Advertising	400,000	19%
Social Media Marketing	300,000	14%
Display Advertising	160,000	9%
Pay-Per-Click Advertising	100,000	4%
Agency Fees	100,000	4%
Email/Direct Marketing	50,000	3%
Marketing Programs	40,000	3%
Public Relations	100,000	6%
Total	$1,965,000	100%

Before: Some of the information in the right matrix repeats what's found in the left matrix. Also, right matrix info is poorly formatted and arranged, making it visually out of sync with what's in the other matrix.

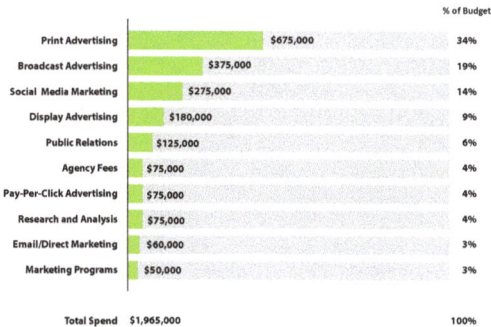

BUDGET SUMMARY

		% of Budget
Print Advertising	$675,000	34%
Broadcast Advertising	$375,000	19%
Social Media Marketing	$275,000	14%
Display Advertising	$180,000	9%
Public Relations	$125,000	6%
Agency Fees	$75,000	4%
Pay-Per-Click Advertising	$75,000	4%
Research and Analysis	$75,000	4%
Email/Direct Marketing	$60,000	3%
Marketing Programs	$50,000	3%
Total Spend	$1,965,000	100%

After: A first glance at the "Before" slide suggests that two slides are needed — one for each matrix. But simply adding the "Percent of Budget" data to the information in the left matrix eliminates redundancy and the need for two slides.

CORVETTE: EIGHT GENERATIONS

C1 1953-1962

C3 1968-1982
Longest Generation
15 model runs

C6 2005-2013
ZR-1 MUSCLE

C7 2014-2019
LAST FRONT ENGINE

C4 1984-1996

C8 2020-PRESENT

GM introduces American
auto industry icon

1983: THE
Missing Corvette
Model Year

C2 1963-1967

C5 1997-2004

THOROUGHLY
MODERN

STING RAY

60 YEARS AND COUNTING!

GM

Before: Slide is too "busy" – information overload. Thumbnail photos are too small. Title is in all caps (the written equivalent of shouting). GM logo is not needed.

Corvette: Eight Generations
C1   1953-1962

- Built in Flint, MI, then St. Louis
- Fiberglass body
- Solid-axle suspension
- Drum brakes
- Small-block V-8 replaces underperforming 6-cylinder
- All convertibles

After (Slide 1 of 8): Creating multiple slides probably is a better option, especially if the presenter plans to discuss each Corvette generation in some detail. This also keeps the audience from reading ahead. Using a larger Corvette photo from each generation adds visual impact. Unreadable footer information has been eliminated. Yet another approach would be to turn that "before" slide into a "build" – adding info for each of the eight generations one click at a time.

Slide Design: Some Rules of the Road

Logos: Avoid putting your company's logo on every slide. Limit its use to your first and last slides. If your company mandates a logo on all slides, place it in the lower right corner. A logo should not compete (in size or placement) with the primary message on the slide.

Photos: They need to be large enough so the audience can see the content. Avoid thumbnails. Also, don't use too many photos on a single slide. Photo montages (especially multiple headshots arranged in a grid) are cliches – overused and generally ineffective. Instead, opt for one compelling, memorable image.

Footers: A footer is a line or block of text at the bottom of some slides. It contains information about something on the slide – author, source, date, copyright, etc. These annoying and usually unreadable "footnotes" in small type are better placed in your handouts. Keep them off the slides you project.

White Space: White space is any section of a slide that's unused. Resist the temptation to fill every part of your slide with text, images or ornamentation. Give your content room to breathe. Your audience will benefit. At the same time, too much unused space (especially if it's actually white) can be jarring. Strike a balance.

Text: Choose a simple, readable font (typeface). *Avoid script fonts* and **decorative fonts**. Sans serif fonts (those that do not have extending features called "serifs" at the end of letters) work best for projected slides. Text in this book is set in sans serif type.

Deciding what font size to use depends on projection screen size and its distance from the audience. But a good rule of thumb is that titles should be 36-44 pt. (point) and all other text should be at least 24-28 pt. A presenter should never utter these words: "I know you really can't read what's on this slide, but let me tell you what it says." Make sure the font colors contrast with the slide's background.

Also, generally avoid complete sentences. No unnecessary words: think: billboard wording. Be careful of spelling and grammar. Make sure capitalization (upper- and lower-case letters) is consistent. If the slide contains math, is it correct?

Multiple Images: Don't assemble multiple charts, graphs or blocks of text on a single slide — even if the data are related. Use a separate slide for each element, or depending on the amount of information, "build" the slide — one item at a time.

The Presentation Genius of Steve Jobs

Steve Jobs was an amazing presenter. His product introductions for Apple were legendary. People attended them as much to see Jobs perform, as to see the new products Apple had created. (Watch a few launches on YouTube.) Here are some observations about what made him such a great communicator:

He could sustain audience attention. The average attention span of most adults during a business presentation is about twenty minutes. Jobs could (and frequently did) hold audience attention for well over an hour, in part because of the way he talked about Apple products.

He used figurative language, including analogies. Analogies are among the most powerful tools of persuasion. And Jobs used lots of them. When introducing the iPod in 2001, he said, "iPod is the size of a deck of cards." While he spoke, a rotating photo of a familiar box of Bicycle-brand playing cards was projected onscreen.

Four years later, he described the iPod Nano this way: "And yet all of this weighs one and a half ounces – 42 grams. That is less than eight quarters in your pocket."

He entertained. One of my favorite Steve Jobs moments: He pulled an iPod out of his pocket and talked about it briefly. Then, pointing to the small, coin pocket on his signature jeans, he said, "Ever wonder what this pocket's for? I've always wondered that. Well, now we know ..." and pulled out an iPod Nano.

Jobs knew how to use Keynote (Apple's version of PowerPoint). Never did you see him "talking to the screen" reading a lot of words. Instead, he'd be looking at the audience while photos or a few key words reinforcing his message appeared on screen. Also, not once did he ever use bullet points or those often-unnecessary agenda slides that begin most presentations.

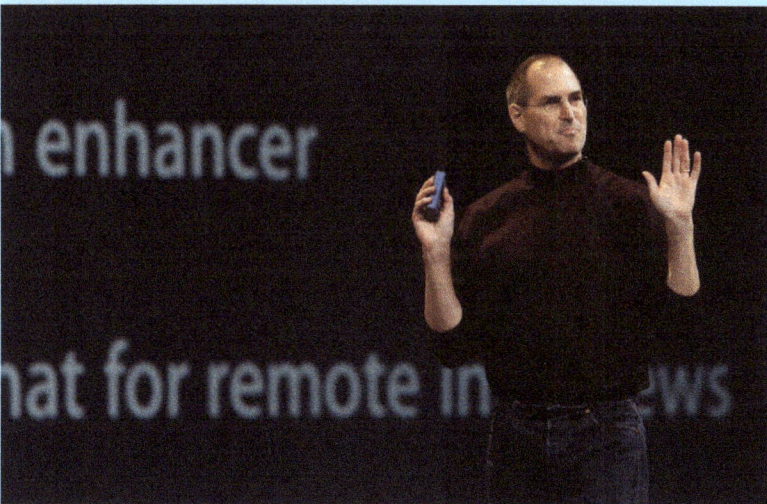

Audiences attended Apple product launches in part to see the "insanely great" presentation skills of Steve Jobs.

Even the best PowerPoint visuals lose their power if they're not well delivered.

Chapter 6

Okay, It's Showtime. Up on Your Feet!

Let There Be Light . . . But Not Too Much

When you present, make sure the room you're using and the equipment in it are properly set up. Don't rely on someone else to do this or even to know what's appropriate. You're the one who will either benefit or suffer from the setup, so consider this your responsibility.

One element worth special attention is lighting. Presenters who use PowerPoint want to be sure the audience can see their slides. Too much room light, especially near the screen, can make that difficult. Today, most conference rooms, lecture halls and the like are designed so you can control the lighting in various parts of the room. Take advantage of this. Turn off the lights near the screen; dim the house lights.

In some rooms, your choice is limited. One switch controls all lights; you can have them either on or off. In this case, opt for the former – even if that means your visuals are harder to see. Why? It's important that the audience be able to see you – not some faceless, shadowy figure. You're as much a part of your message as your slides are. Gestures, facial expressions and eye contact with the audience matter. They provide additional meaning, and the audience needs to see them.

Remember what you did when you wanted your kids to fall asleep? You read to them in a darkened bedroom. Avoid presenting in a totally dark room.

Need another reason to avoid presenting in a totally dark room? Remember what you did when you wanted your kids to fall asleep? You read to them in a darkened bedroom.

Face Your Audience, Not the Screen

In most business room setups, PowerPoint visuals are projected on a screen or monitor behind the presenter. That's why we usually see many presenters "talking to the screen" (or worse yet, reading what's on it). The visuals are being used as a prompt. Instead of maximizing eye contact with the audience, the presenter ignores the audience for long periods or stands sideways (an awkward, unprofessional stance) – dividing eye contact between screen and audience.

In Western cultures, eye contact is valued. Most business and social situations call for "involvement." And one of the best ways to show it is through eye contact. When you glance at someone for just 1-2 seconds, your eyes are darting, a habit that undermines credibility. And eye contact that exceeds 10 seconds might be interpreted as intimidation or intimacy. When we're excited and fully engaged, we tend to look at someone for about 5-10 seconds.

Great presenters have mastered this 5- to 10-second, sustained eye contact skill. They deliver one complete thought to one pair of eyes. In large-group situations, they focus on several individuals who serve as proxies for the entire audience. Another skill they've mastered is looking at the audience at least 85 percent of the time while presenting and looking at their notes no more than 15 percent. (I call this the 85/15 rule.)

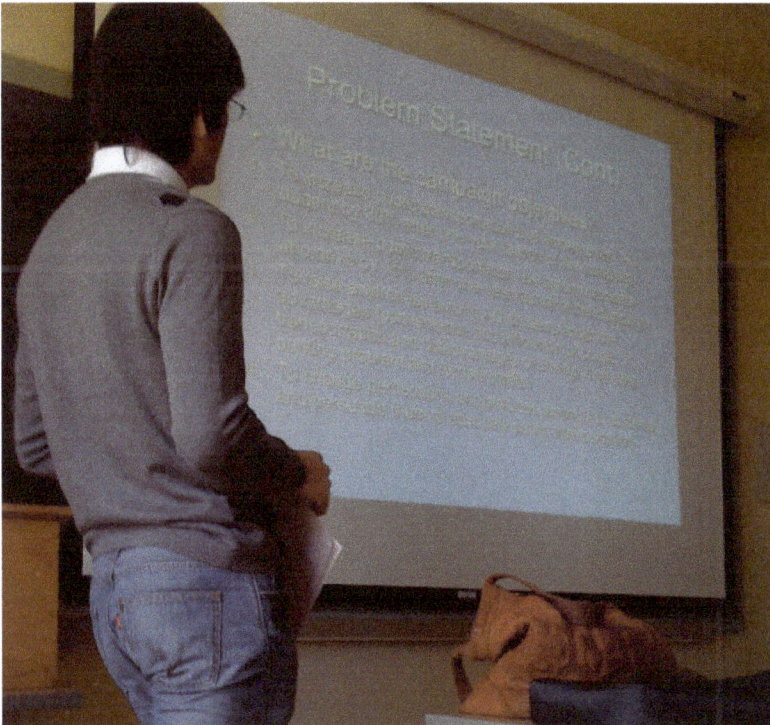

So where are those notes, if not on the screen?

They're on your laptop, which can be placed anywhere – on top of a lectern or table. (Wireless display adapters can connect your laptop to a large screen without using a cable.) This setup enables you to quickly glance down at your notes and then back at the audience.

A less ideal, but workable, option is to have your PowerPoint visuals as hard copies, which can be used just like the script of a speech.

And the least-ideal option is to use the room's large screen as your prompt. In this case, you glance quickly at the screen, but turn back to the audience for a long period. To do this, you must be very familiar with your material. (BTW: It's perfectly fine to turn to the screen when you are guiding the audience through something on a particular visual.)

Weaning yourself off using the screen as your prompt is one of the single most important ways to improve your delivery skills. It may take time, but the results will be worth it. It will set you apart from other presenters.

Use a Remote

Watch most presenters who use PowerPoint and you'll see them standing in one spot tethered to their computer so they can advance their visuals. This prevents the kind of natural movement, including moving around the room, that can help energize an audience.

There are several other ways to advance your visuals. One is to have someone else, maybe a colleague, do it. But then, the audience usually hears you repeatedly utter that annoying phrase, "Next slide." Try to avoid this approach. It weakens your delivery.

Another option is to use PowerPoint's auto advance feature. Yes, you decide how much time elapses between slides, but having a set change pattern can cause you to rush through your presentation or keep you from adequately commenting on slides that deserve that extra bit of attention. Try to avoid this approach as well.

Then there's the remote. Use it.

Remote controls first appeared in World War I. Germany used radio frequency devices to crash some of their naval vessels into Allied ships. In the Second World War, remote controls detonated bombs for the first time.

Today, remotes operate with three different technologies: Infrared remotes (IR) use light to carry signals between the remote and the device it's controlling. Most TV and stereo remotes use IR technology. However, these remotes work at a limited range of about 30 feet, and you must point them directly at the receiving device.

By contrast, RF (radio frequency) remotes use radio waves instead of light waves. Garage door openers, car alarm fobs, radio-controlled toys and most presentation remotes use this technology. It works up to 100 feet from the receiver.

Bluetooth is another option.

If you deliver digital presentations, here's some advice:
- Invest in a remote. RF remotes are affordable and come in many varieties. Be careful of choosing one that's too large and has too many bells and whistles. What you need is something that quickly activates the following four essential features: forward, reverse, black screen and laser pointer. Some remotes also incorporate a timer, so you can see how long you've been speaking.
- The best remotes are small and ergonomically designed to fit in your hand. They're barely visible to the audience. (That's a good thing.)

- Don't draw audience attention to your remote by pointing it at the screen. (With RF remotes, it's not necessary.) You want the audience to focus on you and what you're saying, not on some piece of equipment in your hand. If you're an occasional presenter, make sure you're familiar with how the remote you're using works. Skillful use of a remote can contribute favorably to the "performance aspect" of your delivery.
- Test the remote before every presentation. This is especially important if you're linking it to equipment that's not your own.
- Have extra batteries for the remote.

What about That "B" Key?

One of the least-known features of PowerPoint is what happens when you hit the 'B' key.

One of the least-known (and least-used) features of PowerPoint and Keynote is what happens when you hit the "B" key. Hit it, or better yet, hit its equivalent on most remotes and the room screen will go black.

Think about the following situations:

1. Your title slide is up on the screen and the audience has had time to read it. As you walk

confidently up to the lectern, you hit a button on your remote (no one in the audience even notices) and the screen goes dark. Then you begin your remarks – maybe a self-introduction, maybe a powerful story. Regardless of what opening you use, audience attention is focused exclusively on you and what you have to say – not on what's up on the screen. A bit theatrical? You bet, but what's wrong with that?

2. It's the middle of your presentation and you just showed a bar graph recapping three consecutive quarterly declines in sales and revenues for your firm. Ouch! There needs to be an immediate detour – a discussion of why this happened and what can be done about it. So you pause your presentation, black the screen and prepare for some uncomfortable questions and comments. If needed, that bar graph is queued up, available at the push of a button.

3. You've come to the end of your presentation. You just finished discussing the information on the final content slide. Now it's time for Q&A. You black the screen to prevent that last slide from becoming a distraction during what could be a lengthy Q&A segment. A lot of presenters fail to do this, never realizing that throughout the Q&A segment, the audience will keep looking at and thinking about the image that's still onscreen.

Effective presenters use a variety of skills. Among them: controlling the flow of information. Using the "B" key helps you manage audience attention. When you're ready to return to the screen, hit the "B" key again.

Two final suggestions: If you hit the "W" key while in PowerPoint, you'll get a white screen. Avoid this option; that plain, bright white screen can be jarring. Also, some conference room A-V control systems allow you to black the screen through a "mute video" button. But that approach can be a bit more complicated — requiring you to move back to the lectern to access that button. Stick with the quick, single press of the "black screen" button on your remote.

Laser Pointers: Use with Caution

Let's take a trip down memory lane. You're back in grade school. Atop the blackboard which runs nearly the entire width of the classroom is a strip containing the alphabet — both upper- and lower- case ABCs in cursive writing. The teacher, holding a yard-long, wooden pointer with a black rubber tip, points to each letter and asks the class to recite the alphabet. Remember those wooden pointers? (BTW: As a Catholic grade-school student, I saw nuns also use that pointer across the knuckles of misbehaving kids!)

Years later, you're in a conference room delivering a business presentation. You step up to a wall map, pull out a metal, pen-sized telescoping pointer and use it to show your colleagues the route your company will use to move crude oil from one part of the world to another.

Those two types of pointers, while still around, are no longer widely used. They've been replaced by laser pointers. A laser pointer or laser pen is a small, handheld device that emits a low-level laser beam. You use it to highlight something of interest by illuminating it with a small, bright spot of colored light — usually red.

Like those telescoping pointers (mentioned above) that most presenters ended up "playing with" (i.e., repeatedly expanding and retracting them during their presentation), laser pointers are often misused.

One problem probably has to do with the fact that because most laser pointers are incorporated into a remote, they are immediately and easily accessible. So presenters feel compelled to overuse them. That's why you see presenters circling words or phrases on screen as they use them. It's totally unnecessary, and in fact it's a distraction. Don't do it.

Remember, the purpose of the laser is to draw audience attention to something important. If there's a complicated

Not sure when to use a laser pointer? Maybe this will help:

You've probably seen television sports and weather broadcasters diagram and analyze a sports play or weather pattern. They do this by using a telestrator – a device that allows them to sketch something over a moving or still video image on a touchscreen. That overlayed image ends up on your TV screen. Think: Troy Aikman, Tony Romo or some other football broadcast commentator drawing squiggly lines on an instant replay image. (Football great John Madden was an early adopter of the telestrator and had great fun using it.)

Some football plays can be hard to follow or understand, so the commentator uses the telestrator to draw attention to something specific that occurred on the field and explain it. Do likewise with your presentation. Decide whether an image is especially important or might be difficult for the audience to understand, and if so, use your laser pointer to help them make sense of it.

Not every football play needs detailed analysis. And not every element in your presentation needs laser pointer treatment. Use the tool sparingly.

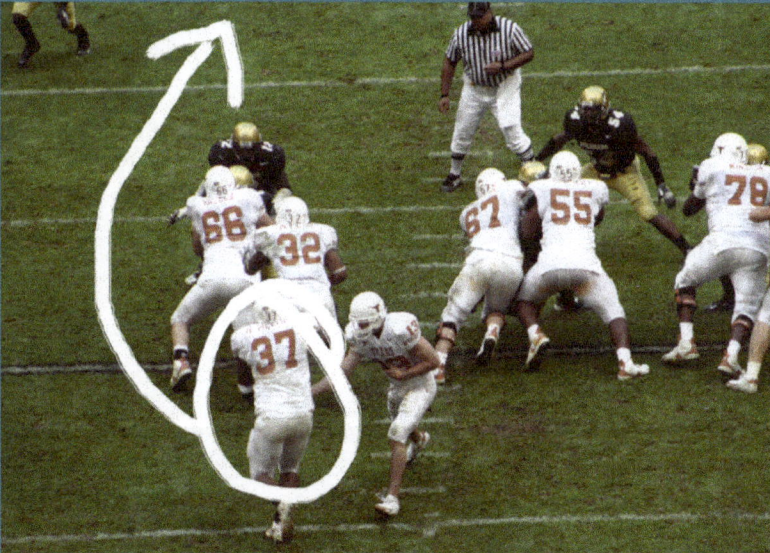

flow diagram or numerical table on screen, using a laser pointer is a good way to help the audience focus its attention appropriately. Or if there's a photo containing an element that's especially important, you can direct audience attention to it to make sure people don't miss something. In other words, use it sparingly. Treat the laser pointer as you would a colored marker when you're reading – have a good reason to underline something on the page.

Another potential problem worth monitoring: the small size and low power of the laser beam make it invisible until it hits a surface. Make sure ambient room light is appropriate and that there's enough contrast between the dot of light and the color of your on-screen images.

And finally, what about safety concerns with lasers? They're real. Even low-power laser beams can cause serious eye damage if the light happens to hit someone's gaze for several seconds. Fortunately, the eye's blink reflex makes this highly unlikely. Exercise caution.

Plan B

Presentation technology, while highly reliable, is not failure proof: Projector bulbs burn out. PowerPoint freezes. Internet connections fail. Computers crash.

Because presentations today rely so heavily on technology, savvy presenters leave nothing to chance. They identify potential problems and have a game plan to deal with them. Here are a few ideas on what to do when things go wrong:

- If a technical problem surfaces, acknowledge it immediately and directly with the audience. (Don't keep people in the dark like airline pilots and gate crews do with passengers when there's a flight delay.) Tell them you're trying to fix the glitch or ask if anyone in the audience can help.

Most audiences will give you about ten minutes to correct the problem; after that, they will get restless.

- If possible, take an unscheduled break. Then use that time to troubleshoot the problem.
- Many venues such as conference centers and hotels, and even some company facilities have A-V technicians on site and on call. Before you present, get the name and phone or text number of that individual. While the technician is addressing the problem, take that unscheduled break or have something specific set aside to discuss that relates to your topic. It's similar to the appendix in a book. Q&A is another possible option, especially if you were able to deliver at least some of your presentation.
- If your audience is fairly small (e.g., just a few people), have and distribute hard copies of your slides (printed prior to the presentation) or quickly email or text them the PowerPoint.
- With large groups, encourage the audience to access your PowerPoint on their laptops, tablets or smart phones. (Who doesn't have one of those devices with them?) They can do this if you've uploaded your slides onto a special link on your company's website or onto a sharing platform such as Dropbox. (Dropbox does not require audience members to have a Dropbox subscription – just a password which you'll provide for them to access the visuals.) In short, there's a way to give large audiences quick access to your visuals, assuming a Wi-Fi connection is available where you are.
- If the technical problem cannot be resolved in a timely fashion, move forward with your presentation if it can be delivered effectively without visuals. (Tell the audience they'll get the

visuals later.) Perhaps it's a slightly different and shorter version that you deliver, but one that still hits the mark. Don't be afraid to try this if you know your topic well, can think quickly on your feet or have invested some time beforehand to map out an alternate version of your remarks. (Talk about being prepared!) By soldiering on and nailing the content, you'll score points with the audience. . . and be remembered. . . favorably.

Slides, Notes and Handouts: Not One and the Same

For most presenters, their PowerPoint slides, the notes they use to deliver their presentation, and the handout they provide to the audience are one and the same. In other words, the on-screen visuals, which usually contain lots of detail, also serve as notes or a prompt for the presenter. And those same visuals become the hard-copy or electronic handout for the audience. This "one size fits all" approach is usually a mistake. If you've read this far in the book, you probably know why.

Effective use of presentation visuals involves the following:

- On-screen slides are there to support, not be, the presentation. Their purpose is similar to that of billboards or headlines – to get audience attention and provide a clear, succinct message that can be processed in a few seconds. The presenter is there to fill in the details. If your slides are loaded with detail, why bother to present in person? Just send the audience a copy of your slides instead. (Sarcasm!)
- The detail you need to stay on track to deliver your key messages should be in your notes. PowerPoint

software developers recognized that most presenters could use some help remembering what to say about each slide. So they introduced Notes in 2003.

Notes are essentially prompts you add to each slide. They don't appear on screen, so the audience can't see them. They're displayed separately on your monitor for your eyes only.

As you create each slide, a box appears, saying, "Click to add notes." Type in what you want to say about the slide. It can be a few words or a detailed script. (Opt for the former.) When you're presenting, your slides will appear on screen, and your laptop or monitor will show the current slide and the notes for it. You'll also see a preview of the upcoming slide. Another option is to print out your notes.

Some words of caution: When working with clients who are struggling when they present, I frequently ask to see their notes. What I usually find is too much detail, including complete sentences – often single-spaced and in small type. (What's worse are handwritten notes.) The best notes are a few words and phrases. They will keep you from reading to the audience or struggling to find the info you need among the clutter. Once again, solid knowledge of your topic along with practice and a few "trigger" words as notes will serve you well.

- The proper place for all that detail most presenters put on their projected slides is the handout (whether hard copy or electronic). This usually means that you need to create two versions of your PowerPoint – one for the screen and one for the leave-behind. Yes, that means a bit more work, but if your goal is to communicate effectively (recall the difference between a production and communication objective), doing a bit more work is worth it.

Some questions about PowerPoint you've probably asked yourself.

Chapter 7

FAQ

Q: What's the ideal number of slides for a PowerPoint presentation?

A: Ask that question of Siri or do a Google search, and here's what the "experts" will tell you: For a 15-minute presentation: 7 slides . . . or 15 . . . maybe 20-30, and for a 20-minute presentation: 10. Confused? The answer to the question is simple: use as many slides as you need to support your content. Some presenters cram too much info on each slide to minimize the total number of slides. Worry less about the number and more about whether those slides are needed, and if so, are properly formatted and used. For example, if your presentation includes a tour of some historic home, you could have quite a few exterior and interior photo slides. But you might project them quickly, so the number of slides is immaterial.

Q: How long should a slide remain on screen?

A: Long enough for the audience to process it. We've all been frustrated by presenters who move too quickly from one slide to the next – before we've had time to digest the information. Give careful thought to the kind and amount of info on each slide when determining how long you'll show it. Ignore anyone who tells you there's an ideal SPM (slides per minute) number. The best rule of thumb is to have one idea per slide. Also, keep in mind that today's audiences (especially younger ones) are used to quick scene changes in movies, music videos, commercials. And TV viewers are using their remote-control devices to change channels between 36 and 107 times per hour. So, a faster-paced, presentation delivery style can work.

Q: Is there a way to keep an audience from reading ahead of where I am on my slides?

A: One problem with "word" or "text" slides is that most people can read at 600 words per minute (WPM), while the speaker is probably talking at 150-200 WPM. You're on your first bullet point, but the audience may already be reading bullet number three. One way to maintain control is to use a "build" or "reveal" where you sequentially reveal information on the slide. You add each bullet one at a time — allowing you to discuss it without the audience being able to read ahead. This is similar to what presenters did years ago with overhead projectors; they used a sheet of paper to cover up portions of the transparency until they were ready to reveal the information. Use a reveal — but only if you're going to spend some time talking about each element in the sequence. Otherwise, you'll end up hitting the forward button at rapid-fire speed.

Q: I'm a PowerPoint newbie. What's the best way to learn how to use this software?

A: PowerPoint might appear intimidating, but it's user friendly. Many users learn by doing. But if you need some help, plenty of resources are available.

Books: There are 7,000 books about PowerPoint listed on Amazon! (Many are out of date and some are out of print.) Nearly all of them deal primarily with how to create slides. Two of the most popular books come from Microsoft and from the "for dummies" brand.

Tutorials: There are plenty of online tutorials (YouTube and written). Also, many libraries offer beginner and advanced PowerPoint workshops – often free of charge. Community colleges and municipal adult education programs are other resources.

Q: Are there companies out there that can create slides for me?

A: Absolutely. Quite a few actually. These firms employ designers, copywriters and graphic artists who can improve or redesign an existing presentation or create one from scratch. But their services aren't cheap. For example, one firm charges $11 to clean up your slides, $43 for a complete makeover and $66 to create a slide deck from information you provide. That's per slide! And those are "starting at" prices for a 3-day turnaround. Faster service is more expensive. You can find these companies on the internet. Another option is to work with freelancers . . . and a lot of creative types (both talented and not-so-talented) have hung out their PowerPoint shingles. Find these independents through companies like Upwork that specialize in freelance workers.

Q: Are there any presentations I can watch for tips on using PowerPoint?

A: Chances are you've heard of something called a TED Talk. TED began in 1984 as a conference focused on Technology, Entertainment and Design. Today, TED Talks are given by thought leaders in a variety of fields. More than 4,000 of these talks are archived online at www.ted.com. Not all TED talkers use PowerPoint, but many of them do. And because TED speakers get some coaching (e.g., length of talk, avoiding "commercial" messages, proper use of visuals, etc.), you'll see some pretty impressive performances. Some of the most-watched talks are flagged.

Q: If I provide the audience with copies of my slides, should I do so before or after the presentation?

A: Ideally, slides or notes should be given after a presentation – to keep the audience from thumbing through the material while you're speaking. However, this is not always practical. For example, in some meetings or events, your visuals may already be part of a packet of materials given to attendees when they arrive. Or, if the topic is technical or highly complex (e.g., analyst presentations), the audience may need the material while you're presenting. In these situations, your presentation delivery skills must be stellar so the audience will choose to listen to you rather than read the material.

Reminder: today, most people prefer to get handouts electronically. And there are apps available that let you communicate with attendees before, during and after your event. Also, don't ignore the power of sending an additional brief text or email several days after a meeting recapping your key messages. This is impressive and effective repetition and reinforcement.

Q: What about using PowerPoint in virtual presentations? Any suggestions?

A: Presenting virtually requires some extra planning and resourcefulness to engage an audience. Younger generations especially will be simultaneously web surfing, checking or sending emails and text messages, even playing games. This impolite behavior has become normal. The good news: multi-tasking and distractions occur only when interest wanes. So, your presentation needs to be more interesting than any distraction.

Among the ways PowerPoint can help you get audience attention is by provoking emotion through powerful images, video, animations and audio files. PowerPoint lets you insert all of those to create a multi-media experience.

Slides loaded with text are especially deadly in virtual presentations. Use them and an audience will tune you out.

And how do you hold that attention once you get it? Build in change: Make sure PowerPoint is not the only thing to appear full screen for the entire presentation. At times, project PowerPoint full screen. At other times, have yourself appear full screen. And still other times, project PowerPoint along with a thumbnail image of yourself.

When you do appear on screen, keep this in mind: You may have noticed that the built-in camera on most laptops sitting on a desk is below the user's eye level. The result? An unflattering view looking up at the user's nose (or the room's ceiling). Place your laptop on a stand or on a stack of books or boxes and raise the camera until it's level with the top of your head, then tilt the screen down.

Other suggestions for on-screen presence:

Posture: While sitting, lean forward (it increases energy and shows engagement) with hands on the desk in front of you. Avoid hands under the desk or on the edge (weak, tentative, unengaged). Have most of your forearms on the desk, but not elbows.

Gestures: Hand movements are more pronounced on screen than they are in person because you are framed in a limited area. Keep your hands from getting too close to the camera (they appear larger than they are in proportion to the rest of your image).

Eye contact: Your camera is your surrogate audience. Over time, you'll get comfortable looking at this inanimate object, but don't constantly stare at it. Occasionally, look at your notes, glance down (never up), or look at your screen.

Q: In terms of effective use of PowerPoint, I'm buying what you're selling. But I'm not sure my company feels the same way. Any advice?

A: Some of the ideas in this book run counter to what many companies are used to or even mandate when it comes to business presentations. One of my clients is a Japanese auto manufacturer operating in the Unites States. American managers at the plant tell me that their Japanese bosses consider presenters who use too few slides or use slides with too little information, to be poor or incomplete thinkers. So, these managers intentionally use ineffective visuals. I get it. Some advice:

- Don't do anything that might jeopardize your career.
- Recognize that we're in a global world and that there are cultural differences that impact communication. For example, western cultures value direct eye contact; not so in Asian or Middle Eastern cultures. Those Japanese auto execs? Keep in mind that English is not their first language, so those "cluttered" PowerPoint slides may have helped them better understand what was being discussed. In short, keep your audience in mind when developing and delivering your presentations.
- Companies too have cultures. And those cultures can impact such things as how employees are expected to present. Be flexible; take some of my suggestions with a grain of salt. Guidelines should not be hard and fast.
- Walk before you run. Make incremental changes. Rather than deciding not to use PowerPoint in a particular presentation, instead, use fewer slides than you normally would. Rather than put three charts on a single slide, display those charts on

three separate slides. Keep the text slides, but eliminate complete sentences and unnecessary words. These kinds of changes are subtle, non-threatening to would-be critics, and may even help change some minds. Effecting change in PowerPoint use is a long-term process.

- Detail-laden slides are standard fare in most companies. If your company or boss wants "everything <u>and</u> the kitchen sink," deliver it in the leave-behind, but develop a shorter version for the screen.
- Become an advocate for presentations that differ from the norm. Why not see if your company's training or communications function will arrange a lunch-and-learn or a workshop on effective presentations. Most people in business are hungry to improve the design and delivery of their presentations.

Abandon old habits, adopt new ones, and create and deliver a PowerPoint presentation that dazzles your audience.

Chapter 8

Parting Thoughts

The "fight or flight" response so helpful to our earliest ancestors is alive and well – ready to help audiences tune in or tune out to your presentation.

Not to end on a downer, but the deck is stacked against every PowerPoint presenter. Here's why:

Think back to our earliest ancestors. One of the things we know about them is that they had smaller, underdeveloped brains. Scientists use the terms, "first," "primitive," "reptile" or "crocodile" brain. It was not designed for reasoning; its reasoning power was quite limited. Instead, it focused on survival. It enabled these ancestors to effectively manage threats. For instance, if a dangerous animal approached, this primitive brain triggered an immediate "fight or flight" response. To be sure, our ancestors were well served by this brain.

Today, our brains are larger and more developed. In the neocortex, for example, reside the amazing skills of language, logic and creativity. But modern humans still have vestiges of that primitive brain. Incoming information is being received and screened there before it can move on to other parts of the brain.

Now, most of us don't encounter threatening animals very often, if at all. But that primitive brain is ready to protect us from other modern-day threats – such as boring presentations. Think about the mindset of most people before they attend a business presentation: "Boy, I'm really busy; I wish I didn't have to go to this thing." "Not another PowerPoint presentation!" "I bet this is going to be boring." "I already know this stuff."

Your audience comes predisposed to fear or resist your presentation. They've been burned in the past by what passes for acceptable presentations.

Which brings us to why the deck is stacked against every presenter: Your audience comes predisposed to fear or resist your presentation. They've been burned in the past by what passes for acceptable presentations. That primitive part of their brain is ready to trigger the fight or flight response. "Flight" could mean not coming, leaving early, tuning out, multi-tasking. You don't want that. You want "fight" – meaning, the audience is fully engaged – ready and eager to listen.

Okay, one final comment before we're done: In this book, I've taken some shots at PowerPoint . . . and how it's being used. I've tried not to pull any punches. So let me make amends. The truth is PowerPoint is a very popular, unpopular tool. It's likely you've sat through it — more than once. Maybe tolerated it. Probably used it. But now you know how to use it to treat your audience to the kind of presentation they rarely encounter but might always remember. So, abandon old habits, adopt new ones, and create and deliver a PowerPoint presentation that dazzles your audience and gets them to listen to, hear, understand and act on what you say. It's easier than you think.

Notes

Introduction

Shortly after taking the reins: Louis V. Gerstner, Jr., *Who Says Elephants Can't Dance? Inside IBM's Historic Turnaround*, (New York, Harper Collins, 2003).

The Wall Street Journal reports: June Kronholz, "Even Second Graders Use PowerPoint in Classrooms," *The Wall Street Journal*, November 12, 2002.

Chapter 1

Back in 1986, researchers from the University of Minnesota: Douglas R. Vogel, Gary W. Dickson, John A. Lehman, "Persuasion and the Role of Visual Presentation Support: The UM/3M Study," Management Information Systems Research Center, School of Management, University of Minnesota, Minneapolis, Minnesota, June 1986.

One interesting take on visual communication: Gary A. Williams and Robert B. Miller, "Change the Way You Persuade," *Harvard Business Review*, May 2002.

The idea for PowerPoint: Pia Lehner-Mittermaier, "The History and Evolution of PowerPoint," SlideLizard Blog, April 20, 2020.

Chapter 2

Remember the 1980s movie Gung Ho: *Gung Ho*, Paramount Pictures, 1986.

In the movie Apollo 13: *Apollo 13*, Universal Pictures, 1995.

In the film *Field of Dreams*: *Field of Dreams*, Universal City Studios, Inc., 1989.

At the 2008 Macworld Conference and Expo: Carmine Gallo, *The Presentation Secrets of Steve Jobs*, (New York, McGraw-Hill, 2010).

In a TED Talk on malaria: Bill Gates, "Mosquitos, malaria and education," TED, February 2009, (www.ted.com).

Chapter 3

One more point about using visuals: Roy Underhill, *Khrushchev's Shoe and Other Ways to Captivate Audiences of 1 to 1,000*, (New York, Basic Books, 2000).

For example, in the documentary, *An Inconvenient Truth*: *An Inconvenient Truth*, Paramount Classics, 2006.

Peter Norvig, a computer scientist and sometime critic: Peter Norvig, "The Gettysburg Powerpoint Presentation," (www.norvig.com).

In 2003, the Space Shuttle Columbia: Tad Simons, "Does PowerPoint Make You Stupid?" *Presentations*, March 2004.

But the board also pointed a finger: Clive Thompson, "PowerPoint Makes You Dumb," *The New York Times*, December 14, 2003.

Nancy Duarte, who wrote a thoughtful book: Nancy Duarte, *slide:ology: The Art and Science of Creating Great Presentations*, (North Sebastopol, California, O'Reilly Media, 2008).

In 2003, Tufte published a booklet: Edward R. Tufte, *The Cognitive Style of PowerPoint: Pitching Out Corrupts Within*, (Cheshire, Connecticut, Graphics Press, 2005).

Amazon founder Jeff Bezos: Madeline Stone, "A 2004 email from Jeff Bezos explains why Powerpoint presentations aren't allowed at Amazon," *Business Insider*, July 28, 2015.

Steve Jobs, CEO, Apple Computer and Pixar Animation Studios: Steve Jobs, Commencement address, Stanford University, Stanford, California, June 12, 2005, (http://news.stanford.edu/).

Former Admiral William H. McRaven: William H. McRaven, "Make Your Bed," Commencement address, University of Texas at Austin, Austin, Texas, May 17, 2014, (https://www.rev.com/blog/transcripts/admiral-william-mcraven-make-your-bed-commencement-speech-transcript).

Michael Ward, Senior Research Fellow, University of Oxford: Michael Ward, "Of Hills and Dales," 2015 commencement address, Hillsdale College, Hillsdale, Michigan, May 9, 2015, *Imprimis*, May/June 2015.

Father Paul D. Scalia, Vicar for Clergy, Diocese of Arlington: Father Paul D. Scalia, Eulogy for Justice Antonin Scalia, Basilica of the National Shrine of the Immaculate Conception, Washington, DC, February 20, 2016, *USA Today*, February 20, 2016, (https://www.usatoday.com/story/news/politics/2016/02/20/transcript-rev-paul-scalias-eulogy-his-father-justice-antonin-scalia/80667122/

John F. Kennedy: John F. Kennedy, Inaugural address, Washington, DC, January 20, 1961, National Archives, (www.archives.gov).

Franklin D. Roosevelt: Franklin D. Roosevelt, Speech to a joint session of Congress, Washington, DC, December 8, 1941, (https://www.owleyes.org/text/pearl-harbor-speech/read/text-of-roosevelts-speech).

Martin Luther King, Jr.: Reverend Martin Luther King, Jr., March on Washington for Jobs and Freedom, Washington, DC, August 28, 1963, (www.rev.com).

OH NO, NOT POWERPOINT! RELAX, HELP IS ON THE WAY.

Who's not familiar with Dr. King's historic: Clarence B. Jones and Stuart Connelly, *Behind the Dream: The Making of the Speech that Transformed a Nation*, (New York, St. Martin's Press, 2011).

Ronald Reagan: President Ronald Reagan, Address to the Nation on the Explosion of the Space Shuttle Challenger, Oval Office at the White House, Washington, DC, January 28, 1986, Ronald Reagan Library & Museum, Simi Valley, California, (Reagan.library@nara.gov).

Chapter 4

Speaking of audience sensitivities: Barbara Bush, "Choices and Change: Your Success as a Family," Delivered at Severance Green, Wellesley College, Wellesley, Massachusetts, June 1, 1990, *Vital Speeches of the Day*, July 1990.

Some well known products began life: Will Hellpern, "11 famous products that were originally intended for a completely different purpose," *Business Insider*, April 1, 2016.

In the James Bond movie *Skyfall*: *Skyfall*, Metro-Goldwyn Mayer Studios Inc. and Danjaq, LLC, 2012.

For example, Thomas Jefferson: Lilyan Wilder, *7 Steps to Fearless Speaking*, (New York, John Wiley & Sons, Inc., 1999).

One year to the day of George Harrison's death: *Concert for George*, November 29, 2002, Warner Music Group.

He did it by delivering a climate change slide show: *An Inconvenient Truth*, Paramount Classics, 2006.

Former U.S. Treasury Secretary Timothy Geithner: "Money, Power and Wall Street," *Frontline*, 2012, WGBH Educational Foundation.

"Work is a rubber ball": Gary Keller with Jay Papasan, *The One Thing: The Surprisingly Simple Truth Behind Extraordinary Results*, (Portland, Oregon, Bard Press, 2013).

Chapter 5

Pie charts are the workhorses of the chart world: Stephen Few, *Show Me the Numbers: Designing Tables and Graphs to Enlighten*, (El Dorado Hills, California, Analytics Press, 2012).

Chapter 6

Remote controls first appeared in World War I: Julia Layton, "How Remote Controls Work," *HowStuffWorks*, February 11, 2021, (https://electronics.howstuffworks.com/remote-control.htm).

Photo Credits

Chapter 1

A Brief History of PowerPoint
Glen LeLievre/The New Yorker Collection/The Cartoon Bank

Chapter 2

Carousel projector
Photo courtesy Kodak

Apollo 13
Apollo 13, Universal Pictures, 1995

Blackboards
Patrick Pahlke on Unsplash

Whiteboards
This is Engineering on Unsplash

Steve Jobs/MacBook Air
Kimberly White via Getty Images

AI-generated mosquito
Sabrina Belle on Pixabay

Lime
Showcake on iStock

Chapter 3

A PowerPoint Parody
Courtesy Peter Norvig

PowerPoint's Role in the Space Shuttle Disaster
Presentations Magazine

Steve Jobs/Stanford University
Jim Gensheimer/*San Jose Mercury News* via Getty
Images

Admiral William H. McRaven
Martha Miller/The University of Texas at Austin

Michael Ward
Imprimis, Hillsdale College

Father Paul D. Scalia
Pool via Getty Images

JFK Inauguration
Army Signal Corps/John F. Kennedy Presidential Museum
& Library

Franklin D. Roosevelt
Pictorial Press Ltd/Alamy Stock Photo

Martin Luther King, Jr.
Associated Press

Ronald Reagan
Courtesy Ronald Reagan Library

Chapter 4

Barbara Bush
Rick Friedman/Pool via CNP

Does Your Presentation Need an Agenda Slide?
GeorgiosArt/iStock by Getty Images

Dodging Bullets
Kubicka on Shutterstock

Concert set list
Courtesy Rolling Stones

Brock Purdy
AP Photo/Godofredo A. Vasquez

Your Final Slide
Summary: Black Salmon on Shutterstock

Dolly Parton: Wes McFee on Unsplash
Glass balls: Tengyart on Unsplash
Businessman bowing: Minerva Studio on Shutterstock

Chapter 5

Volodymyr Zelensky
PA Images/Alamy Stock Photo

Planning to Use Photography?
Light bulb: Public Domain Pictures from Pixabay
Handshake: Gerd Altmann from Pixabay
Globe: Gerd Altmann from Pixabay
Bullseye: Lindsay Jayne from Pixabay
Smiling workers: PikWizard
Pad, pencil, etc.: Arek Socha from Pixabay
Light switch: eric 1513 on iStock

Slide Makeovers
True West Petroleum: curraheeshutter on iStock, e-crow
on iStock

The Presentation Genius of Steve Jobs
Siegle/Alamy

Chapter 6

Laser Pointers; Use with Caution
Wikimedia Commons

Chapter 8
Life_in_a_pixel/Shutterstock

About the Author
Author photo courtesy Corner Suite Communications

About the Author

Ken Haseley is founder and principal of Corner Suite Communications, a firm that helps leaders and aspiring leaders assess and improve one of their most important, yet frequently ignored, skill sets – communicating effectively through the spoken word. His firm specializes in communicating with investors and those who influence them.

A graduate of Kent State University, Ken has a Bachelor of Science degree in communications and government. Additionally, he holds an MBA from the University of Dallas.

His corporate affiliations have included Diamond Shamrock and Occidental Petroleum, where he served as a communications director and worked in investor relations. During 1982, he was on loan to the Reagan Administration in Washington, DC.

His experience as a communications coach and trainer spans some twenty years and involved working with companies and other organizations in twenty countries. Ken has a loyal following among the C-suite executives he coaches.

He has lectured at MIT's Sloan School of Management, served as a visiting professor at Ivanovo State University in Russia, and developed and teaches Communications for Leaders in the Executive MBA Program at the Bauer College of Business at the University of Houston.

Ken's first book, *Change the Way You Communicate: Why You Should. How You Can.*, challenges business leaders to eliminate comfortable, but outdated and self-defeating communication. His articles and speeches have appeared in a variety of publications, including *Fortune*, *Vital Speeches of the Day*, *Investor Relations Update*, *Moscow Times* and the *Houston Business Journal*.

After hours: In addition to business communications, Ken's other interests are sailing his 33-foot "Connecticut Yankee" and driving his '59 and '65 Corvettes. Those cars don't have names, but if they did, they'd be Tod and Buz – a tip of the hat to the two lead characters who traveled the country (in a Corvette) in the '60s TV show *Route 66*.

www.cornersuitecommunications.com
ken@cornersuitecommunications.com

www.ingramcontent.com/pod-product-compliance
Lightning Source LLC
Chambersburg PA
CBHW052126030426
42335CB00025B/3136